THE
TREE
with
MANY
BRANCHES

A Collection
of Essays in
Computational
Phylogenetics

TOMMY RODRIGUEZ

THE TREE WITH MANY BRANCHES
A COLLECTION OF ESSAYS IN COMPUTATIONAL

iUniverse books may be ordered through booksellers or by contacting:

iUniverse
1663 Liberty Drive
Bloomington, IN 47403
www.iuniverse.com
844-349-9409

ISBN: 978-1-6632-0649-7 (sc)
ISBN: 978-1-6632-0648-0 (e)

Print information available on the last page.

iUniverse rev. date: 08/14/2020

THE

TREE

with

MANY
BRANCHES

A Collection
of Essays in
Computational
Phylogenetics

TOMMY RODRIGUEZ

THE TREE WITH MANY BRANCHES
A COLLECTION OF ESSAYS IN COMPUTATIONAL

iUniverse books may be ordered through booksellers or by contacting:

iUniverse
1663 Liberty Drive
Bloomington, IN 47403
www.iuniverse.com
844-349-9409

Because of the dynamic nature of the Internet, any web addresses or links contained in this book may have changed since publication and may no longer be valid. The views expressed in this work are solely those of the author and do not necessarily reflect the views of the publisher, and the publisher hereby disclaims any responsibility for them.

Any people depicted in stock imagery provided by Getty Images are models, and such images are being used for illustrative purposes only. Certain stock imagery © Getty Images.

ISBN: 978-1-6632-0649-7 (sc)
ISBN: 978-1-6632-0648-0 (e)

Print information available on the last page.

iUniverse rev. date: 08/14/2020

CONTENTS

FOREWORD

I n 2003, scientists at the Human Genome Project sequenced the entire 23 chromosome pairs of the human genome for the very first time. This achievement is hailed by many as one of the greatest scientific triumphs in human history. It set the stage for a host of fields, disciplines, and applications in bioscience that would later provide more insight about the way we look at ourselves, and at the vast diversity of living organisms around us. Since then, tens of thousands of species have been sequenced in some fashion; most of which are bacteria. Although the exact number is unknown, publicly accessible genomic databases, such as GenBank and GOLD, now house roughly an estimated 100 million+ records of complete DNA sequences, gene sets, RNA, or protein sequences. The field of bioinformatics was born out of the need to manage, analyze, and examine raw genomic data in meaningful ways. Computer technology and life science now had a place where they could both reside together.

The inner workings of evolutionary biology have always fascinated me. Early on, reading the works of Charles Darwin, Francis Crick, Richard Dawkins, and others, influenced me to pursue the biosciences at the highest levels of academia. Like many undergraduate students, I flip-flopped college majors on a few occasions but mostly stayed within the confines of life science and other computer-related fields. At the time, academic programs in computational biology were scarce. Consolidating my diverse backgrounds was not a feasible option at the universities that I attended. Only later, during my late twenties, was I able to merge both disciplines at the University of Maryland University College, where I earned a graduate degree in biotechnology.

My general specialization was, and still is, bioinformatics (or the application of computer technology to biological data); and it was during this time in graduate school that I also took a particular interest in computational phylogenetics – a field of research that generally falls within the umbrella of bioinformatics. I then spent much of my graduate career writing algorithms for distance-matrix models and computer code for genomic analysis tools. As we will see later, many of the same computational biology programs that I adopted in graduate school were also utilized toward my later works.

One at a time, I filled the years that followed graduation engaged in several different research projects of my own. Essentially, this book is a compilation of those studies. I found it somewhat liberating in having the flexibility of independent research. Of course, the downside to working independently also lies in the limitations of self-funding and in the lack of scholarly collaborations. Still, independent research has given me opportunities to explore a wide array of inquiries and topics in phylogenetics that may not be present otherwise. Opting not to work or intern for a private firm where research might be dictated by senior staff members, today I submit and publish my work through various internationally well-known peer-reviewed open access journals. More recently, I have expanded my studies to include ecological restoration for a path toward conservation genetics.

To start, I would not presume to describe my research as novel or breakthrough or even cutting-edge. However, it is in the application of computational techniques, experiment design, and probabilistic models where my research finds a stronghold. As a matter of practicality, the original manuscripts have been edited for a broader audience due to its highly technical language. The essays compiled in these pages have undergone a facelift, from their original scientific format into a more reader-friendly layout, as to better accommodate two different perspectives – both experts and non-expert alike.

I thought it best to begin the book by outlining my standard practices, procedures, methods, and techniques for building phylogenies, as means to educate the reader on how I later reach inferences or conclusions. While this book is a recollection of essays in computational phylogenetics, the central theme deals directly with the inferences brought forth in conjunction with the evolutionary relatedness of organisms and groups of organisms in phylogenetic context. Topics range from the evolutionary implications of biological aging to cancer-associated trends in primate populations; others include the influences of radiation-induced evolution to the origins of domesticated dog breeds. The first case study presented in Chapter 3 involves my work investigating the evolutionary development of aminoglycoside resistance genes in pathogenic bacteria, which was also featured in another book entitled, *Top 10 Contributions in Bioinformatics & Systems Biology Vol. 2018* (Avid Science). One essay also includes a fully comprehensive sampling of a near-complete mammalian phylogeny based on complete mitochondrial biomarkers.

Alas, this book also contains a bonus chapter. Some might even consider it controversial. As the latter years of graduate school were winding down, I spent some time preparing myself to become a life science educator and *evolution denial* is a topic that repeatedly resurfaced in my curriculum. The thought might seem silly to future generations as they read this entry, but in this modern time there is still much controversy surrounding the topic of biological evolution in many sectors across America. As a researcher and life science educator, I find it hard to remain silent in the face of opposition to a scientific fact. The final chapter is a more simplified version of an essay written ten years ago that discusses this topic in detail. Embodying wide areas of indisputable evidence, it lays out a compelling case against opposing viewpoints with the purpose of correcting any misconceptions about the theory of biological evolution.

To Ethan and Roxanne;
My love, my life, my joy.

1

WANT TO BUILD A TREE?

When Charles Darwin was on board the HMS Beagle in 1837, he had a game-changing suspicion about living systems. Species were not immutable, he thought. Instead, *they* (populations of organisms) descend and diversify from lineal predecessors. He wrote down some of his first ideas involving common ancestry on what we now know as the Darwin Notebooks (from the Voyage of the Beagle). [1] In his notes he depicted the famous "I think" sketch, where he attempts to illustrate patterns of diversification from nested lineages in the form of connected and dispersing branches. Without knowing the importance that it would later signify, Charles Darwin unwittingly sketched the first phylogenetic tree.

Darwin's tree did not have a root nor did it represent any specific set of actual lineages or groups of organisms (that we know of). His tree was abstract and totally conceptual in nature. Today, most phylogenetic trees are illustrated as cladograms; or rooted trees containing nested hierarchies of relatedness between living organisms. During his time, Darwin had no prior knowledge of the mechanisms, or even the existence of, a cell. Less much so did he know about genetics, which only came about at the turn of the twentieth century with the discovery of DNA by James Watson and Francis Crick. Darwin's theory of evolution by natural selection explains the process in which certain heritable traits and variations help organisms survive and reproduce to become more common in a population over time; [2] but it does not explain the mechanisms by which those variations occur and are later inherited by offspring.

Generally speaking, a phylogenetic tree can be defined as a diagram showing a series of evolutionary pathways from a common ancestor to different descendants. And, it describes these evolutionary pathways via divergent events that split series of nodes into corresponding taxa. There are essentially two lines of data that we examine when building phylogenies: morphology and genetics. For purposes of the material covered in this book, we will look at the latter. To greater or lesser extents, all organisms share molecular variation patterns due entirely to their common ancestry to one another. Through molecular sequencing, different degrees of relatedness can be measured and determined with high scopes of confidence.

In phylogeny research, two types of trees are commonly used in practice: rooted and unrooted trees. A rooted tree is a tree in which one of the nodes is stipulated to be the root, and the direction of ancestral relationships is determined in a nested hierarchy of outgroup and ingroup lineages. As you move from the root to the tips (or taxa), you are moving forward in time. [3] The nodes located on a rooted tree are considered a point of *speciation* (or the event where populations diverge into new distinct species). Unrooted trees can also be useful in showing general relatedness between and among organisms, but do not necessarily imply an ancestral root. Edge lengths (also known as branches) are important features of phylogenetic trees, as they can be interpreted as rough time estimates.

Building a phylogenetic tree from genomic data requires the application of several different computational techniques, which are covered in the following sections. In any such case, the very first step in any tree building exercise starts with a well-thought-out experiment design. This is when we think about the topic that we are interested in investigating, and essentially, how it will be executed. Next, the researcher must decide on sequence selection, followed by data collection and sequence analysis. Lastly, the algorithms one chooses toward sequence alignment and distance-matrix modeling is a final, critical step for reconstructing a phylogeny, where utilizing the right combination of methods can result in better resolution and overall accuracy.

Genomic Databases

With these general principles guiding us forward, we are ready to build a phylogenetic tree. So, let us get to it then. If we were researchers building a phylogenetic tree as part of an experiment, we would start by compiling a genomic dataset of organisms to arrange into a tree. Suppose you are interested in finding the degree of relatedness between a group of pathogenic bacteria. Perhaps your focus is aimed at identifying the evolutionary pathways among old-world monkeys that led to modern humans. Whether you want to explore one biological inquiry or another – from the very small to the moderately large – much of the data that you will need to conduct an experiment is at your fingertips (literally, and figuratively speaking). Today, most genomic data can be accessed and collected via publicly accessible databases.

As noted earlier, two lines of data can be used to build a phylogenetic tree. However, we can expect to get much higher resolution utilizing genome-scale data to infer phylogenies, rather than just physical traits alone (although, combining both may be the best practice if the corresponding data is available). This is mainly due to the amount of information we can get from sequence comparisons. When comparing genomic sequences, there is a general rule of thumb: a larger number of differences corresponds to less related species, whereas a smaller number of differences corresponds to a more closely related type. Furthermore, selecting the right sequence dataset is an important factor in the experiment design stage, as it can have consequences on the overall accuracy of a sequence alignment. Sequences themselves are available in many different formats, sizes, and types.

Methods, Procedures & Algorithmic Selection

In molecular phylogenetics, experimentalists seeking advice on their experimental design are mainly confronted with "what can go wrong when using a certain method" and "what are the most important factors influencing accuracy." [4] From the perspectives of methodology, I often utilize a specific combination of statistical algorithms due to their continued reliability: (1) Kalign for multiple sequence alignment (MSA); (2) pairwise comparison; and (3) PHYLIP neighbor-joining matrix method. An accurate and fast MSA algorithm, Kalign is a dependable algorithmic selection for purposes of obtaining highly robust alignments. [4] Kalign is an extension of Wu-Manber approximate pattern-matching algorithm, based on Levenshtein distances. This strategy enables Kalign to estimate sequence distances faster and more accurately than other popular iterative methods. Kalign has been shown to be about 10 times faster than ClustalW and, depending on the alignment size, up to 50 times faster than other iterative methods. It also delivers better overall resolution. [5]

PHYLIP neighbor-joining matrix modeling can generate highly probable diagrams amid scenarios involving low degrees of variance, regardless of alignment size. Appropriated toward tree-building exercises, PHYLIP neighbor-joining is an accurate and statically consistent polynomial-time algorithm that does not assume that all lineages evolve at the same rate, and it constructs a tree by successive clustering of lineages, setting branch lengths as the lineages join [where a set of n taxa requires $n - 3$ iterations; each step is repeated by $(n - 1) \times (n - 1)$]. [6] This method incorporates a set of default parameters for distance matrix model F84. Additional bootstrapping compilers are often set to 70% – 75%, while transition ratios are generated automatically under default settings. [4] For reference purposes, the following formula demonstrates a standard neighbor-joining Q-matrix algorithm:

$$Q(i,j) = (n-2)\ d(i,j) - \Sigma\ \{n,\ k = 1\}\ d(i,k) - \Sigma\ \{n,\ k = 1\}\ d(j,k)\ (1)$$

Pair to node (distances):

$$(f,u) = \tfrac{1}{2}\ d(f,g) + \tfrac{1}{2}(n-2)\ [\Sigma\ \{n,\ k = 1\}$$
$$d(f,k) - \Sigma\ \{n,\ k = 1\}\ d(g,k)]\ (2)$$

Taxa to node (distances):

$$d(u,k) = \tfrac{1}{2}\ [d(f,k) + d(g,k) - d(f,g)]\ (3)$$

Summary

With proper training and stark curiosity, the application computational phylogenetics can lead us to a better understanding of the complexities involved in the diversity and distribution of living organisms over the span of evolutionary time. Phylogenetic trees provide a glimpse into those evolutionary pathways. However, let us not forget that phylogenies themselves are often regarded as inferences to a hypothesis about an evolutionary inquiry. This makes technological perspectives even more important to the tree building process. The higher our degree of confidence in the accuracy of a cladogram, the more reliable our inferences are about the results. In the next chapter, we start with a simple comparison. To begin our journey into the world of computational phylogenetics, I utilize a common genome type to determine the accuracy among various sets of computational algorithms commonly used in phylogenetics. The speed and accuracy of three popular algorithms will be benchmarked, as they can help us better decide on the most reliable combination of methods.

2

TECHNOLOGICAL PERSPECTIVES IN COMPUTATIONAL PHYLOGENETICS

Originally published in the Journal of Advances in Bioscience and Biotechnology

Biological information is compiled of huge amounts of raw data. Collecting, processing, and managing that biological data can be a challenge. Since the turn of the century, modern technology has allowed advanced next-generation sequencing to be achieved with ever-increasing precision. As of 2020, studies in computational phylogenetics largely involve genomic datasets that are manageable through a network of simplified computer systems. Supercomputers are generally not required in most cases, even amongst the most extraneous scenarios. Nonetheless, in consideration of computing for bioinformatics, the following computational factors should always be considered prior to experiment design regardless of the logistics behind sequence data: 1) operating system; 2) processor; 3) physical memory; 4) disk storage; 5) algorithm selection; 6) bioinformatics platform; 7) related software; and 8) networking peripheral(s). Other biometric hardware components may be required.

Early in my academic career, I sought experiments such as these, to compare the accuracy and execution times of various popular Bayesian algorithms for multiple sequence alignment. The results of the experiment later described in this chapter would lean in the direction of one particularly fast and efficient MSA model. This study included complete sequences of mtDNA that had been collected from the NCBI Nucleotide database and imported into a bioinformatics software called UGENE. A series of multiple sequence alignment tasks were performed on thirteen mtDNA sequences of mammalian origin; each sequence represents a species within a unique taxonomical group. As a first option, I selected Kalign for multiple sequence alignment. Several studies have shown significantly large discrepancies in execution times, accuracy, and resolution when Kalign is compared against other computational Bayesian algorithms. In this respective test trial, Kalign for MSA yielded regular timeframes of t > 136.15 s and t < 139.95 s, on five separate instances (Figure 1); far superior than MUSCLE, which required exceedingly longer timeframes per interval.

MAFFT is another speedy alternative for MSA. As illustrated in Figure 1, MAFFT generated remarkable execution times that are comparable, if not better, to the timeframes produced by Kalign. Yet, in cases involving large-scale genomic datasets with increased evolutionary distances, Kalign provides better overall resolution. [4] As the creators of Kalign point out (Lassmann and Sonnhammer), the quality of methods in test sets, namely ClustalW, MUSCLE, and MAFFT, decreased in their own test trials when the number of input sequences was increased. [5] This too became evident as I increased the number of taxonomical groups to my working base-pair alignments. Here, MUSCLE and MAFFT generated a handful of cladogram irregularities that were inconsistent with earlier results containing a reduced volume of sequences.

Before proceeding, I should briefly note that the original FASTA data files did not exceed 220 kb. FASTA format is often used in computational phylogenetics. Light-weight datasets are critical also, among other variables that help reduce potential bottlenecks. Using our genomic datasets to examine computational performance, I was able to quantify the amount of useful workload compared against time and resources. Here again, my results would favor Kalign for multiple sequence alignment. Taken as a whole, Kalign requires minimal amounts of resources for execution, in the shortest amount of time.

In my test trials, I repeated this procedure for multiple intervals until an average mark was obtained. During runtime, CPU frequency levels peaked at 28.8% and hovered between 26% - 28%. At first, the physical memory usage averaged between 25 - 26 MB, and steadily fluctuated during MSA runtime but did not exceed 26.9 MB. Only 1.2 MB of additional RAM was required to perform MSA on any given instance. In comparison, both MUSCLE and MAFFT far exceeded a computationally efficient mark for physical memory usage, as reflected in Figure 2. Figure 3 also highlights the average range for CPU frequency between the three algorithms, where MUSCLE and MAFFT exceed the average mark set by Kalign.

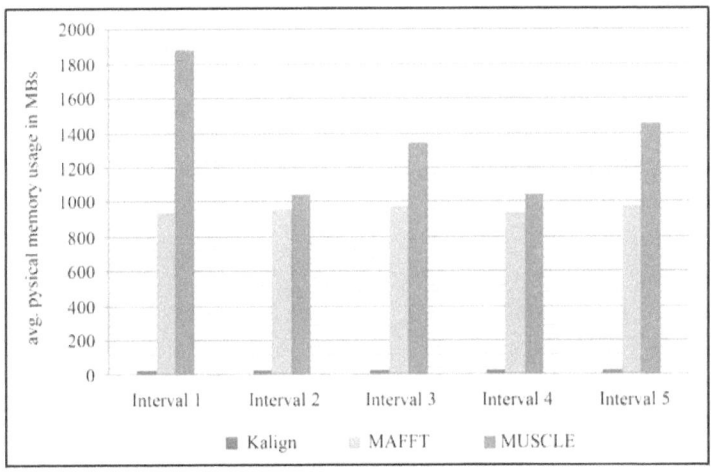

Figure 1. Execution time comparison for multiple sequence alignment: Kalign, MAFFT, and MUSCLE.

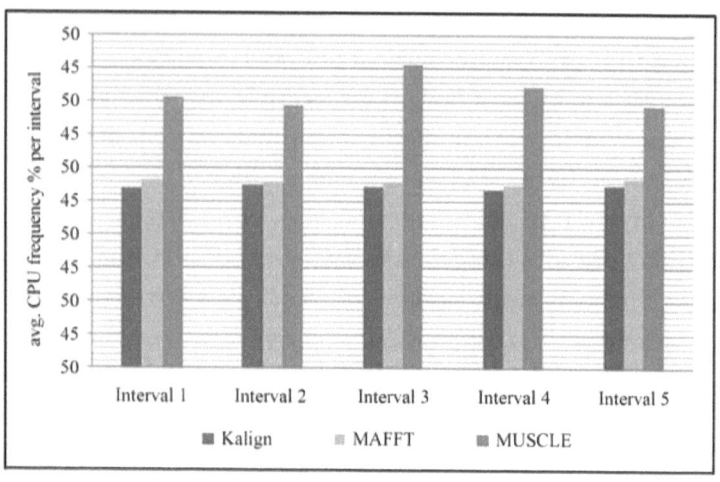

Figure 2. Average CPU frequency comparison for multiple sequence alignment: Kalign, MAFFT, and MUSCLE.

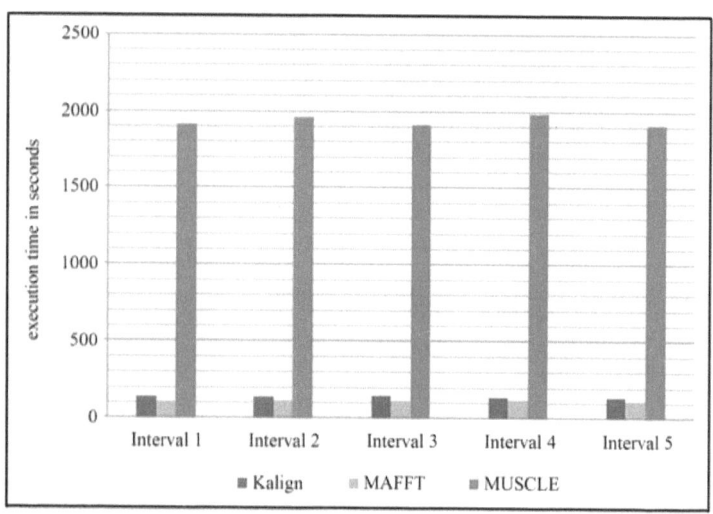

Figure 3. Average physical memory comparison for multiple sequence alignment: Kalign, MAFFT, and MUSCLE.

Utilizing Complete Mitochondrial Genomes in Computational Phylogenetics

For several reasons that are well known to the field of molecular phylogenetics, mtDNA is a suitable choice for examining divergence events among closely related species. The rapid evolution rates in mtDNA create more molecular variance among immediate populations. This has notable advantages when studying ancestral relationships whose divergence times are thought to be no greater than 8 to 10 Myr.[7] Moreover, mtDNA is easier to isolate, purify, and sequence than entire sequences of nuclear DNA (nDNA). Each sample cell can contain a thousand copies of mtDNA, and only a single copy of nDNA. Another advantage is that mtDNA degrades slower than nDNA and it contains a higher prevalence in fossilized remains, which allows genetic comparisons of extinct species and closely related non-extinct species.

Mitochondria is inherited solely through the maternal line (with a few rare exceptions), and it has an important role in phylogeny research. Admittedly, mtDNA alignments are not feasible options for all facets of molecular phylogenetics, especially when comparing molecular variation patterns among organisms that span large evolutionary scales. Such scenarios may produce irregular results. Consequently, a set of issues arise from using matrilineal lineages to build phylogenetic trees: 1) rapid rates in base-pair substitution creates saturation that can result in homoplasy;[7] 2) should male and female history differ in a species, then this marker would not reflect the history of the species as a whole but that of the female portion;[7] 3) hybridization can cause mtDNA to move freely between species and may infer incorrect relationships when building phylogenies.[8] Inaccurate conclusions on several peer-reviewed studies have raised contention about using mtDNA alone in phylogeny research; including a recent paper on the phylogenies of polar bears and brown bears, which resulted in incorrect evolutionary inferences due to hybridization.[9]

Of course, examples of natural hybridization leading to speciation are exceedingly rare, especially in mammals. While most known cases of hybrid speciation occur in plants, the majority of documented instances involving eukaryotes have been observed in fish and insects. [10] As it relates to the phylogenetic experiment described in this chapter, other studies also examining the evolutionary lineages of mammalians, particularly *Elephantidae* and *Sirenia*, have confirmed those relationships via genetics, where some of the most direct support comes in the form of mtDNA. In fact, the results of my own mitochondria-based cladogram support many of inferences raised by others. To start, the phylogenetic tree on Figure 4 illustrates a parent clade containing divergence a series of divergent events that are consistent with the historical record of the family *Elephantidae*. Note the clade containing *Elephas* and *Mammuthus* – sister taxa to *Loxodonata* – extending outward toward *Mammut American*. It then follows that molecular analyses combined with comparative morphology, put manatees and dugongs among the closest living relatives of modern elephants. [11] [12] Again, these evolutionary relationships are further depicted on Figure 4.

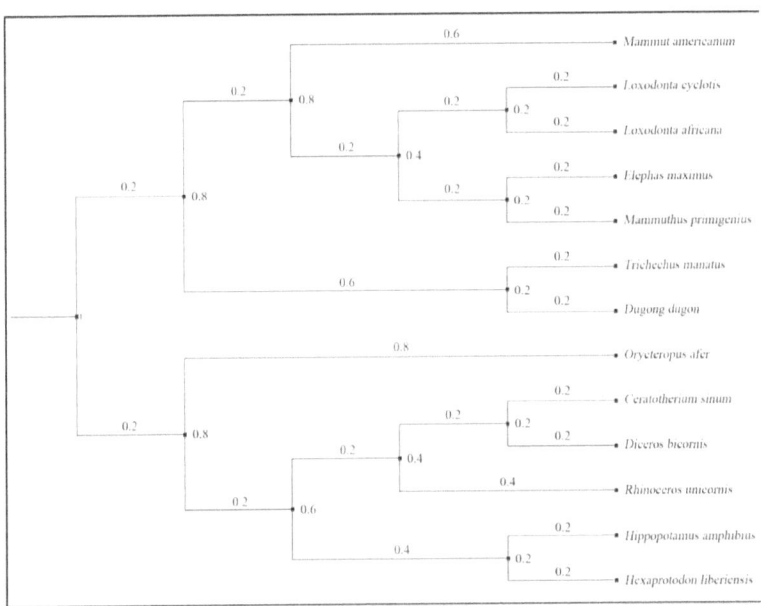

Figure 4. Resulting Phylogenetic Tree (in no particular order): (a) *Loxodonta cyclotis*; (b) *Loxodonta africana*; (c) *Elephas maximus*; (d) *Mammuthus primigenius*; (e) *Mammut americanum*; (f) *Trichechus manatus*; (g) *Dugong dugon*; (h) *Orycteropodidae ofer*; (i) *Ceratotherium simum*; (j) *Diceros bicornis*; (k) *Rhinoceros unicornis*; (l) *Hippopotamus amphibius*; (m) *Hexaprotodon*.

I expanded group taxonomy further as to include *Hippopotamidae, Rhinocerotidae,* and *Orycteropodidae.* It was the Goodman 1981 study which first identified a set of unique molecular similarities in amino acid sequences of α-crystallin A among the aardvark (*Orycteropodidae ofer*), paenungulates manatee, hyrax, and elephant; and these evolutionary relationships have since been detailed by other well-known sources, namely Honeycutt in 2008 and in Nishihara 2005. [11][13][14][15] *Orycteropodidae* shares a fairly high degree of genetic similarity with three distinct groups: 1) *Elephantidae*; 2) *Sirenia*; and 3) *Rhinocerotidae*; whereas *Hippopotamida* are more closely related to modern day cetaceans

13

and contain the least degree of genetic similarity among the six groups in this experiment. Morphological data coupled together with fossil evidence has also shown patterns of close relatedness between four of the six clades. [16] Once again, the full extent of these evolutionary relationships matches the lineal record illustrated on Figure 4.

Summary

Although a handful of inaccurate conclusions have raised questions about its reliability, many researchers, including myself, would still agree that mitochondrial genomes can provide sufficient resolution for reconstructing a robust phylogeny and facilitate the molecular dating of divergence events within a phylogeny; even among time-extended lineages. [16][17] Some experts will argue that mtDNA sequences are only useful for species-level and genus-level analysis. [17] Yet, the boundary that separates populations based on those genomic markers are still not well defined and it should be explored further.

Because computational performance is tied to instances that have potential bearings on the outcome of a phylogenetic experiment, I argue in favor of a practical and simplified approach to phylogeny research. Light-weight genomic datasets, such mtDNA sequences, combined with efficient Bayesian algorithms for computational phylogenetics can help reduce potential bottlenecks, make up for lackluster hardware, and narrow the scope of error. Although this framework is not new to the field of computational phylogenetics, this chapter reinforces its reliability in both performance and accuracy.

CASE STUDIES

3

PHYLOGENETIC CONSIDERATIONS IN THE EVOLUTIONARY DEVELOPMENT OF AMINOGLYCOSIDE RESISTANCE GENES IN PATHOGENIC BACTERIA

Originally published in the Journal of Phylogenetics & Evolutionary Biology

Antibiotic resistance in pathogenic bacteria has been the source of concern in recent times. Each year in the United States, at least 2 million people become infected with bacteria that are resistant to antibiotics and at least 23,000 people die each year as a direct result of these infections. [18] Repeated usage of antibiotic drugs can cause resistance to become more prevalent. Susceptible bacteria are killed or inhibited by an antibiotic, resulting in a selective pressure for the survival of resistant strains. [19] Moreover, resistance is rapidly expanding to include several critical antimicrobials used to treat the most invasive infections.

Today, new findings suggest that antibiotic resistance appeared long before the introduction of antibiotic drugs. Over three hundred sets of homologous protein coding genes for antimicrobial resistance have been identified among the five bacteria types. [20] Interestingly, some of the highest degrees of genetic similarity in antimicrobial resistance genes are shared between bacterial pathogens and modern soil-dwelling varieties. Nitrogen-fixing bacteria are thought to have developed resistance from selective pressures in soil, which acts as a reservoir for antimicrobial resistance. [20] Such a scenario presumes that pathogenic bacteria may have acquired resistance through evolutionary events with a common ancestor of a soil-dwelling bacterium.

This chapter revisits antibiotic resistance as a source of evolutionary development in pathogenic bacteria. By taking a molecular phylogenetic approach to this inquiry, I seek to find homologous correlations in antimicrobial resistant gene families across a broad spectrum of bacteria, as to identify the possible acquisition of those genes through divergent events in evolutionary context. The scope of my investigation will again feature techniques in computational phylogenetics for reconstructing a phylogeny based on two distinct sets of multiple sequence alignments involving antimicrobial resistance genes.

Antimicrobial resistance gene families

The ARDB (Antibiotic Resistance Genes Database) lists approximately three hundred seventy-three protein coding genes for antimicrobial resistance. [21] A significant percentage of those genes are associated with pathogenic bacteria. This chapter concerns itself with one group, of one variety: aminoglycoside resistance genes. Aminoglycoside resistance genes are widely spread in bacteria genera, and they play an important role in antibiotic drug resistance. These genes are characterized by three primary mechanisms of resistance, namely ribosome alteration, decreased permeability, and inactivation of the antibiotics by modifying enzymes. [22]

Antimicrobial resistance spreads as bacteria themselves move from place to place. For decades, soil ecologists have speculated that soil acts as a reservoir for antimicrobial resistance. [20][23] Over time, nitrogen-fixing bacteria have evolved the ability to become antimicrobial resistance as a countermeasure to naturally occurring environmental threats, such as the compounds frequently produced by competing microbes. As preliminary data indicates, pathogenic and nitrogen-fixing bacteria possess a similar genetic basis for resistance but do not share an obvious means for transfer among themselves. A 2012 paper entitled, *"The shared antibiotic resistome of soil bacteria and human pathogens,"* elaborates on the significance of sequence similarities across different bacteria species that occur in a host of different environments. [20][24] In this study, Forsberg demonstrated a high degree of matching DNA sequences between soil-dwelling and pathogenic bacteria, and it provided evidence for the exchange of antimicrobial resistance genes between environmental bacteria and clinical pathogens. [24]

To further support these findings, a more recent study shows that antimicrobial resistance genes found in the bacterial flora of humans must have also developed prior to synthetic and semi-synthetic antibiotics. The study identified several antimicrobial resistance genes in the bacterial flora of humans that are targeted at natural antibiotics of the sort produced by soil microbes. [25] Indeed, I too hold the viewpoint shared among others: soil-dwelling bacteria may be the original source of antibiotic resistance in bacterial pathogens. To test this hypothesis, I examined the various degrees of phylogenetic relatedness for aminoglycoside resistance genes among a broad spectrum of bacteria that occur in different environments.

Gene selection

This investigation utilized two partial sets of aminoglycoside resistance genes (aadA1 and aadA2) for comparative analysis. Aminoglycoside resistance genes are ideal candidates, as they encompass a broad antimicrobial spectrum shared between diverse populations of bacteria. [26][27] These gene families are also generally associated with an exceptionally high-level of resistance to antibiotics. The mechanisms that modify aminoglycosides by adenylylation in Aminoglycoside O-nucleotidylyltra are most notably known to occur in response to antibiotic complex produced by *Streptomyces kanamyceticus* from soil. [27] [28]

The bacterial species appropriated for this study are found to contain resistance strains of these naturally occurring responses. As such, I compiled two distinct FASTA data files containing a combination of eleven nucleotide sequences derived from pathogenic bacteria and soil-dwelling varieties. I ran several BLAST similarity searches against *Salmonella enterica* subspecies strain SRC54, and this procedure returned a significantly high number of homologous sequences to be later used in this study. See Table 1 for accession numbers.

Bacteria Species	Aada1	Aada2
Salmonella enterica	GQ924769.1	DQ836009.1
Escherichia coli	KR028103.1	JQ414042.1
Leclercia adecarboxylata	NG_041647.1*	KM278190.1
Aeromonas hydrophila	KM278193.1	KM278189.1
Riemerella anatipestifer	JF920804.1	AY968682.1
Comamonas testosteroni	KM278197.1	KM278191.1
Citrobacter freundii	JX494725.1	AF175203.1
M.esteraromaticum	KJ575540.1*	KJ575540.1
Laribacter hongkongensis	GU726913.1	GU726907.1
Aeromonas caviae	FM207629.1	KJ568502.1
Providencia stuartii	NC_019375.1*	NC_019375.1*

Table 1. Bacteria identification and sequence data accession numbers.

Sequence analysis and phylogenetic reconstruction

Among the eleven bacterial strains included in each gene family subset, the sequence similarity percentage between them averaged 78.6% and 82.6%, respectively, with values ranging from 54% to 99%. Five pathogenic strains yielded exceptionally high sequence similarity ratios, ranging from 96% to 99% and 98% to 99%, respectively. Based on these estimates, five sequences could be assigned to a subgroup of very closely related strains. It is generally admitted that sequences with greater than 97% identity are typically assigned to the same species, those with >95% identity are typically assigned to the same genus and those with >80% identity are typically assigned to the same phylum. [29] However, due to partial sequence sizes, the latter may not apply here. See Table 2 for sequence similarity ratios.

Species Name	Aada1	Aada2	Sequence Similarity Ratio
Salmonella enterica	0.99	0.99	0.99
Escherichia coli	0.84	0.98	0.91
Leclerciaadecarboxylata	0.78	0.99	0.885
Aeromonashydrophila	0.73	0.99	0.86
Riemerellaanatipestifer	0.96	0.88	0.92
Comamonastestosteroni	0.74	0.99	0.865
Citrobacterfreundii	0.79	0.54	0.665
M.esteraromaticum	0.69	0.69	0.69
Laribacterhongkongensis	0.54	0.68	0.61
Aeromonascaviae	0.81	0.57	0.69
Providenciastuartii	0.78	0.79	0.785
Standard Deviation	0.786	0.826	0.806

Table 2. aadA1 and aadA2 sequence similarity ratios.

Subsequently, one might then project a subgroup consisting of five highly homologous sequences to dictate the trajectory of clade positioning within each tree, beginning with its inner node(s) and extending outward. These patterns highlighted the lineage disbursements in each diagram, where five of eleven highly homologous sequences fell within the closest proximity of all sequence candidates. As Figure 5 illustrates, the taxon represented by the innermost node(s) are assigned to species of clinical pathogens, collectively; whereas, strains positioned among the tree outgroups, occur in diverse environments (sewage, soil and water); including one exclusively soil dwelling strain (*Comamonas testosteroni*).

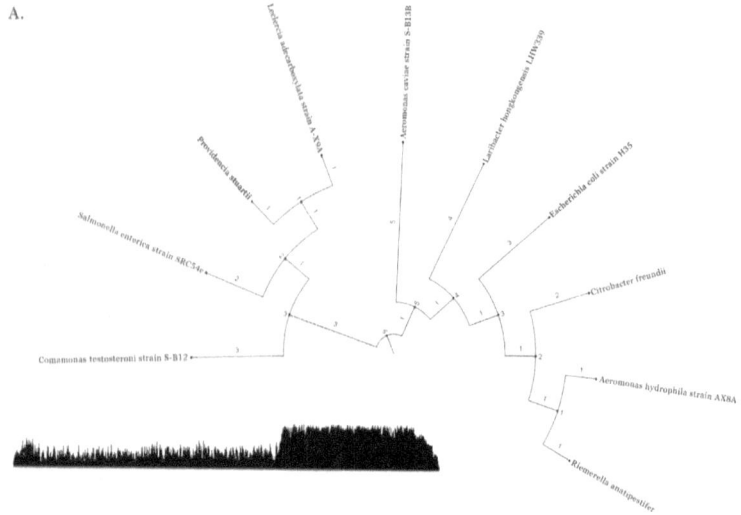

Figure 5. (A) Phylogenetic reconstruction of aadA1 aminoglycoside resistance genes. Taxa order (inner-node to outer-node arrangement): clade a) *Leclercia adecarboxylata* A-X9A, *Providencia stuartii*, *Salmonella enterica* SRC54e, *Comamonas testosteroni* S-B12; clade b) *Riemerella anatipestifer*, Aeromonas hydrophila AXBA, *Citrobacter freundii*, *Escherichia coli* H35, *Laribacter hongkongensis* LHW339, *Aeromonas caviae* S-B13B. Nucleotide substitution area graph included.

B.

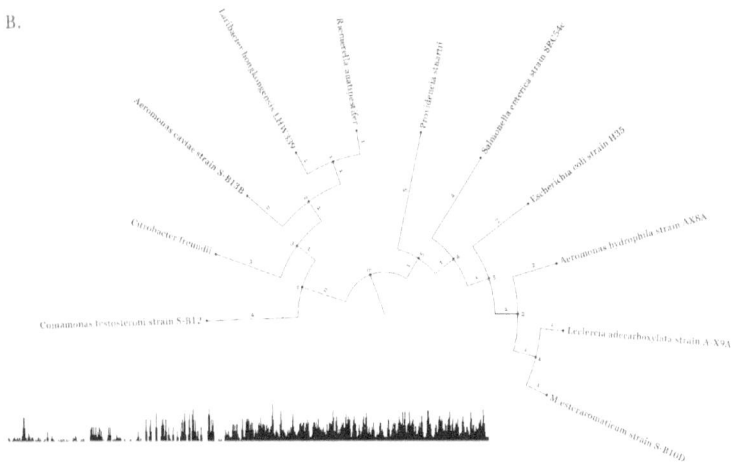

Figure 6. (B) Phylogenetic reconstruction of aadA2 aminoglycoside resistance genes. Taxa order (inner-node to outer-node arrangement): clade a) *Riemerella anatipestifer*, *Laribacter hongkongensis* LHW339, *Aeromonas caviae* S-B13B, *Citrobacter freundii*, *Comamonas testosteroni* S-B12; clade b) *M. esteraromaticum* S-B10D, *Leclercia adecarboxylata* A-X9A, *Aeromonas hydrophila* AXBA, *Escherichia coli* H35, *Salmonella enterica* SRC54e, *Providencia stuartii*. Nucleotide substitution area graph included.

Results

The analysis of aadA1 sequences reveal rooting inconsistencies with that found in Figure 5. Furthermore, among the sister groups located within the inner-most nodes remain three clinical pathogens that correspond to Figure 5. As we move outward from one external node to the next, the arrangement of taxa becomes less distinguishable. This discrepancy is also featured in the similarity ratios shown above, and the nucleotide substitution rates on the area graphs shown on each cladogram.

Yet, despite the inherit differences between them, an underlying trend was identified: (a) positioning among the inner-most and outermost taxon depicted on each diagram - namely clinical pathogens and soil-dwelling strains, respectively – correlate on both instances; (b) soil-dwelling taxa, represented by their position along the outgroups of each tree, appear having older lineages for aadA1 and aadA2 aminoglycoside resistance genes. And thus, by assessing the units of branch length on both diagrams, where the sequence candidates with higher nucleotide substitution rates reside on the far ends, I find very good support for the precursors of aadA1 and aadA2 aminoglycoside resistance genes in pathogens. The significance of these results also provides evidence for the exchange of antimicrobial resistance genes across different hosts, environments, and geographical origins.

Discussion

Antimicrobial resistance in bacteria is not a modern evolutionary innovation. In fact, antibiotics made from compounds produced by bacteria and fungi have existed long before humans formulated the first antibiotic drugs. In nature, antibiotics can increase selective pressure in a population of bacteria, promoting resistant bacteria and supporting its survival prospects. And, as it often occurs in the medical sector, antibiotic drugs are used too often or incorrectly, which can cause resistance to spread faster than it would in natural settings. [30] For this reason, a focus on identifying the evolutionary events that led to the acquisition of resistance in pathogens could help us better understand the interactions that occur between diverse bacteria across a wide range of hosts and environments.

Bacteria use horizontal gene transfer as one primary method for exchanging genetic information. It is also known that recombination plays an important evolutionary role. [31] Although self-inducing genetic mutations in bacterium can create the variation needed within a population to produce new genes for antimicrobial resistance, it is more likely that acquired resistance via DNA transfer between different strains would best explain the homologous correlations observed in antimicrobial resistance gene families. High sequence similarity ratios in aadA1 and aadA2 aminoglycoside resistance genes among distinct species also imply that DNA transfer has occurred between these organisms sometime in the past.

Consequently, the Forsberg study points out that whether shared resistance is confined to genes of mechanisms or applies to many genes with diverse mechanisms of resistance is still unknown. [20][32] Moreover, Forsberg goes on to state, that whether a single horizontal gene transfer event between environment and clinic can result in the de novo acquisition of a multidrug-resistant phenotype is unclear. [32] Thus, looking at my results hereafter, it is difficult to speculate on how likely or unlikely each scenario may be; especially when this investigation did not involve full-length genomic datasets, but only two partial sequences belonging to one gene family from eleven species. In any case, I would simply stress the scope of this study is not to speculate on the mechanisms for acquisition, but rather, to illustrate a phylogeny based on genes for antimicrobial resistance. As such, I have demonstrated that pathogenic antibiotic resistance for aadA1 and aadA2 aminoglycoside resistance genes may have been acquired through evolutionary events with a common ancestor of a soil-dwelling bacterium.

Summary

The exchange of resistance between pathogens and soil-dwelling bacteria emphasizes the clinical importance of the soil resistome. [20][32] From a phylogenetic perspective, this chapter reinforces the inferences already reached by others. Based on my results, I find very good support for the precursors of aadA1 and aadA2 aminoglycoside resistance genes in pathogens. The results also provide evidence for the exchange of aadA1 and aadA2 aminoglycoside resistance genes across different hosts, environments, and geographical origins.

Lastly, it should be noted that a phylogenetic reconstruction involving two partial genomic datasets from eleven distinct species does not substantially improve on the antibiotic resistome as a whole. As others have pointed out, determining the clinical impact of environmental resistance requires a deeper profiling of environmental reservoirs for the organisms and genotypes most likely to exchange resistance with pathogenic varieties. [33] I too propose a more thorough investigation, as to include a wider range of species and antimicrobial resistance gene families.

4

A Glimpse at the Role of Naturally Occurring Radiation as a Contributing Factor to Genetic Variance among Populations of Living Organisms

Originally published in the International Journal of Biology

While the effects of radiation on the cell biology of individual organisms have been widely investigated, there has been a dearth of publications involving the role of naturally occurring radiation in the evolutionary development of living populations. For one, the sun emits short-wave UV radiation that ionizes at about 120 nm to 10 nm, and it can disrupt the normal chemical processes of cells, causing them to become damaged, to grow abnormally or to die. [34] Moreover, radiation can create mutations in DNA that could be inherited over successive generations. [35] [36] [37] [38] Two nucleotide bases in DNA – cytosine and thymine – are most vulnerable to the effects of radiation. [39] For low levels of radiation exposure (or non-ionizing radiation), the biological effects are so small they are thought to cause little or no damage to cells. [40]

Over the course of Earth's history, changes to the atmosphere must have played a key role in the conditions under which life formed and evolved. [41] It is estimated that contemporary organisms would be killed in a matter of seconds if exposed to the full intensity of solar radiation from a primitive Earth. Today, of course, living organisms are protected by an atmospheric ozone layer that effectively absorbs light at short wavelengths. [42] [43] [44] Despite the presence of a modern atmosphere, various known groups of eukaryotic life forms, mostly those belonging to the invertebrate phylum, can tolerate considerable levels of acute ionizing radiation. [45] Other phyla, primarily mammals, are much more hypersensitive in that respect. [46] The range of radioresistance in eukaryotic organisms is considerable and, in some cases, quantifiable. This stark contrast has led me to propose the following inquiry: What role, if any, does naturally occurring radiation play in shaping the outcome of living systems over the extent of evolutionary time? For purposes of this investigation, two distinct orders of eukaryotic organisms that reside on opposite ends of the radioresistance spectrum are examined: scorpions and rodents.

Scorpions are largely resistant to the effects of atmospheric radiation. For instance, their exoskeleton makes it effective at reflecting UV radiation; [47] better shielding its organs from radiation damage and, subsequently, protecting its cells from genetic mutation. *Scorpiones* occur in a vast array of different environments that put it in proximity to the most severe extremes of atmospheric radiation on Earth. [45] A small number of *Scorpiones* species have even been documented to resist doses of radiation as high as 154,000 roentgens. [48] [49] What's more intriguing in this discussion, scorpion morphology has changed relatively little since their first appearance in the fossil record.

Unlike scorpions, rodents are moderate to highly susceptible to the effects of naturally occurring radiation. Ongoing experimentation has shown that rodents have a median lethal dose

(LD$_{50}$) of 7.5 against ionizing radiation exposure, whereas other mammalians range between 3.5 and 12. [50] During the Eocene, rodents evolved and diversified into a wide range of subfamilies that make up about 40% of all mammalian species and they too occur on almost every continent. [51] Not only is their proliferation extensive but also the rates of DNA evolution vary significantly among lineages, which has hindered attempts to reconstruct a robust phylogeny. [52] Most importantly, several studies have largely shown that an increased rate of molecular evolution in rodents is found to be entirely the consequence of a higher mutation rate as compared to other species, families, and genera. [53] [54] [55]

Genetic mutation is the raw material needed for evolution to occur. Natural selection acts upon genetic variation to create the diversity around us. Whether by direct exposure to or by indirect selective pressures from, this chapter explores the possible consequences that naturally occurring radiation might impose on the variance and composition of population genetics. Furthermore, it is proposed that naturally occurring radiation may be implicated in creating genetic variance among different populations of living organisms over extended evolutionary time. To evaluate strength of this hypothesis, a simple statistical experiment is conducted involving 49 cytochrome gene sequences examined in phylogenetic context, pairwise dissimilarity ratios, and molecular clock estimates.

Pairwise Dissimilarity Ratios

After each sequence alignment was consolidated during the course of this experiment, a series of pairwise dissimilarity datasets were calculated using the total length of each cytochrome sequence between any two organisms and divided by the length of the total genomes of the organism in row. The total ratios are then used as a basis for the intensity of genomic discrepancy (or variance) among organisms of the same order. Specifically, for each measurement taken, percentages of pairwise dissimilarity were computed, a broader median within homologous sequences determined, and those figures were then compared against each respective group, lineage, or order.

The results obtained from each dataset are shown in Table 3 and Table 4. Of the global measurements taken, the order *Rodentia* contained the highest total dissimilarity ratio, falling within the range of 9% to 25%; and a median score of 84%. Meanwhile, order *Scorpiones* ranged between 1% and 9% with a median score of 98%. Measurements of *Rodentia* also show far wider concentrations of diffused similarity nearest the middle of a bell curve and disperses in opposite directions relative to the mean. *Rodentia* as a group produced more total variance and those results are visually represented on a dot plot distribution shown on Figure 3. Each raw set (*Rodentia, Scorpiones*) generated a mean of 84, 98 and standard deviation of 4.2648, 2.4162.

Annotation Number / Species / Base Pair	Consensus Percentage
>AB109397.1 *Apodemus draco* (1140 bp)	1.00
>KX790791.1 *Mus cookii* (1144 bp)	0.85
>AF159396.1 *Mus poschiavinus* (1144 bp)	0.85
>FR751074.1 *Mus cypriacus* (1140 bp)	0.85
>AB819920.1 *Mus musculus* (1140 bp)	0.85
>AF520637.1 *Mus caroli* (1191 bp)	0.81
>AY057808.1 *Mus macedonicus* (1145 bp)	0.85
>AF159397.1 *Mus spicilegus* (1144 bp)	0.85
>AB033700.1 *Mus spretus* (1140 bp)	0.86
>AJ698872.1 *Mus famulus* (1140 bp)	0.85
>AB125779.1 *Mus fragilicauda* (1140 bp)	0.87
>AB125777.1 *Mus terricolor* (1140 bp)	0.86
>KY587424.1 *Mus booduga* (1143 bp)	0.86

>AJ233955.1 *Acomys minous* (1141 bp)	0.80
>AJ233957.1 *Acomys cilicicus* (1141 bp)	0.80
>Z96053.1 *Acomys cahirinus* (1141 bp)	0.80
>EF187792.1 *Acomys wilsoni* (1141 bp)	0.80
>EU349734.1 *Apodemus semotus* (1136 bp)	0.91
>FR775869.1 *Rattus andamanensis* (1143 bp)	0.85
>JQ823422.1 *Rattus rattus* (1143 bp)	0.84
>DQ191487.1 *Rattus praetor* (1137 bp)	0.83
>JQ793903.1 *Rattus tanezumi sladeni* (1137 bp)	0.84
>KJ592784.1 *Bandicota savilei* (1139 bp)	0.82
>U87525.1 *Heterocephalus glaber* (1122 bp)	**0.75**
>KY753976.1 *Diplothrix legata* (1125 bp)	0.83
>KJ592782.1 *Bandicota indica* (1139 bp)	0.82
>KY587421.1 *Bandicota bengalensis* (1140 bp)	0.82
>KC735127.1 *Rattus norvegicus* (1143 bp)	0.83
>Z96068.1 *Acomys spinosissimus* (1141 bp)	0.80
>JN247708.1 *Acomys muzei* (1140 bp)	0.81

Table 3. Pairwise comparison of 31 cytochrome b sequences representing order *Rodentia*.

Annotation Number / Species / Base Pair	Consensus Percentage
>AY156582.1 *Pandinus imperator* (658 bp)	**1.00**
>AY156575.1 *Heterometrus swammerdami* (658 bp)	0.99
>AY156572.1 *Heterometrus fulvipes* (658 bp)	0.99
>AY156574.1 *Heterometrus spinifer* (658 bp)	0.99
>AY156573.1 *Heterometrus laoticus* (658 bp)	0.98
>AY156578.1 *Opistophthalmus carinatus* (658 bp)	0.98
>AY156577.1 *Opistophthalmus capensis* (658 bp)	0.99

>AY156576.1 *Opistophthalmus boehmi* (658 bp)	0.99
>AY156579.1 *Opistophthalmus holmi* (658 bp)	0.98
>KT188295.1 *Scorpio fuscus* (658 bp)	0.98
>KT188319.1 *Scorpio kruglovi* (658 bp)	0.97
>KT188221.1 *Scorpio propinquus* (658 bp)	0.97
>KT188328.1 *Scorpio palmatus* (658 bp)	0.98
>JQ514257.1 *Hadogenes paucidens* (653 bp)	0.93
>JQ514256.1 *Hadrurus arizonensis* (659 bp)	**0.91**

Table 4. Pairwise comparison of 18 cytochrome COI sequences representing order *Scorpiones*.

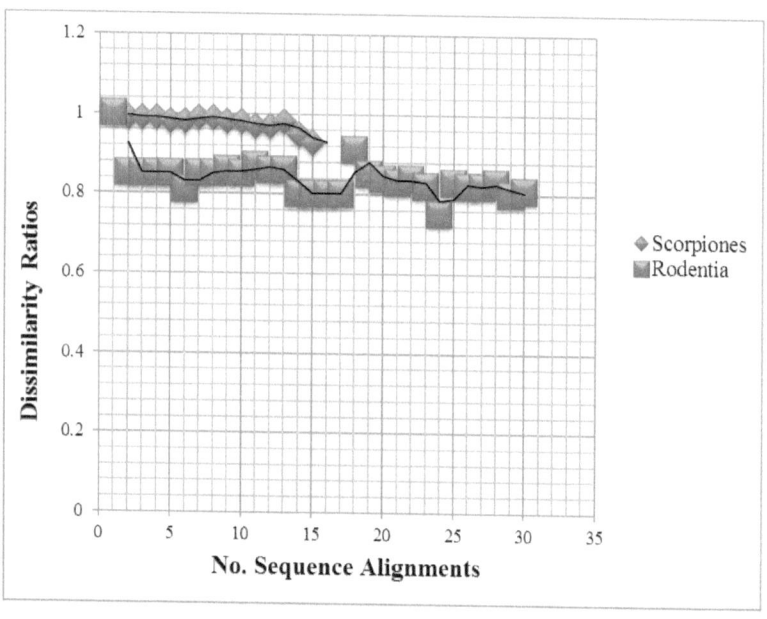

Figure 7. Dot plot distribution of pairwise comparisons between order *Scorpiones* and order *Rodentia*.

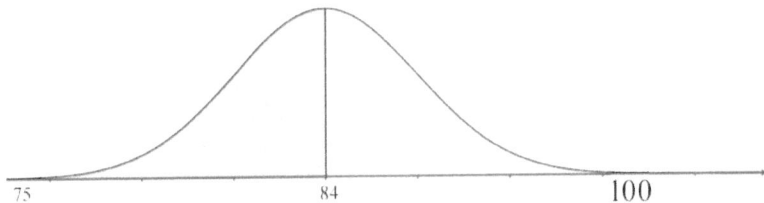

Figure 8. Bell curve distribution (*Rodentia*). Mean: 84, Standard Deviation: 4.2648, Normal Distribution: Pr(X<100) = 0.9819.

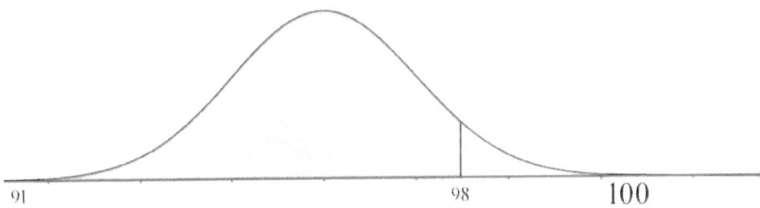

Figure 9. Bell curve distribution (*Scorpiones*). Mean: 98, Standard Deviation: 2.4162, Normal Distribution: Pr(X<100) = 0.7961.

Implications of Evolutionary Trees

Computational phylogenetics can account for evolutionary history and explicitly model trait change along the branches of evolutionary trees. [56] The value of these methods relative to pairwise comparisons has been repeatedly shown in analysis of distance estimations between all possible pairs of sequences in each dataset. Ultimately, the goal of improving distance estimation is to increase the accuracy of the reconstructed tree topology. [57] It then follows that each node estimate on the branch of a tree can represent a degree of confidence about the accuracy of any clade distribution within a given tree. In terms of reliability, bootstrapping is a common method for assessing confidence in the results of phylogenetic analysis. As other sources have noted, bootstrap proportions of ≥70% usually correspond to a probability of ≥95% that the corresponding clade is accurate; [58] [59] and thus, the probability of inferring an accurate clade distribution increase.

The results are simulated under a null hypothesis model. The cladograms derived from the original datasets do not show an excess of larger contrasts for duplication events, nor does it reject the null hypothesis. As shown on Figure 10 and Figure 11, the hierarchical placement of each clade, from outgroup within, are consistent with biogeographical distribution of contemporary species relative to each genus and each order. [60] [61] [62] Additionally, the results agree with recent molecular studies that yielded significant evidence in support of *Rodentia* phylogeny, with genus *Apodemus* and genus *Mus* appearing in the first branches on opposite ends of the outgroup; whilst genus *Rattus* and genus *Heterocephalus* are located among the intergroup clades (Figure 10). [63] Studies done on *Scorpiones* phylogeny are more variable in terms of their agreement on clade placements. [64] [65] Here, both datasets are supported by strong comprehensive bootstrap values (approximately ≥94%) via PHYLIP neighbor-joining method for clade distributions.

Molecular Clock Estimates

The branch length estimates on Figure 11 suggest deep rooted ancestry whereas Figure 10 illustrates several short length clusters that could represent a series of rapid diversification events. This interpretation is based on the units of substitutions per site corresponding to each node on the diagram marked by their descending branch lengths in each lineage. Measures of lineage divergence are not a priori design to be best estimates of time; however, they are often good proxy for time and often interpreted as being measurements of time. [66] To transform these lengths into a time scale, further information is required.

To start, rather than using a relaxed clock model that describes how each branch length *l=r x t* can be decomposed into a rate *r* and a time *t*, evolutionary time scales were postulated by utilizing the distances between neutral substitutions. [67] The neutral substitution rate is often the best estimate we have of an underlying mutation rate. [68] And, deciphering the average mutation rate in any given species should provide further opportunities for estimating species and population divergence times. [69]

Several studies have established the average mammalian genome mutation rate at ±2.2 × 10–9 per base pair per year. [69] [70] [71] Others have examined the null hypotheses of neutral mutation rates among *Arachnida* lineages and have come to some conflicting conclusions. [72] The most accepted mutation rate found in *Scorpiones* is 0.39 ± 0.94 per site × 10^9 years. [73] By combining these external estimates to the results obtained from multiple sequence alignment, phylogenetic analysis, and pairwise comparisons, where *Rodentia* attained several clusters of short-length divergent times and higher rates of neutral substitutions, there is good support for the original inquiry. Patterns of diversification are more frequent among Rodentia lineages compared to the *Scorpiones* order (see Figure 10 & Figure 11).

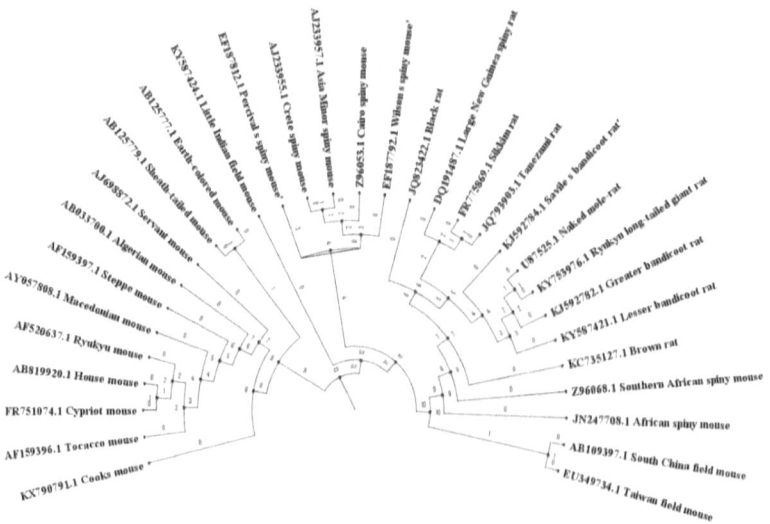

Figure 10. Phylogenetic reconstruction of 31 cytochrome
b sequences representing order *Rodentia*.

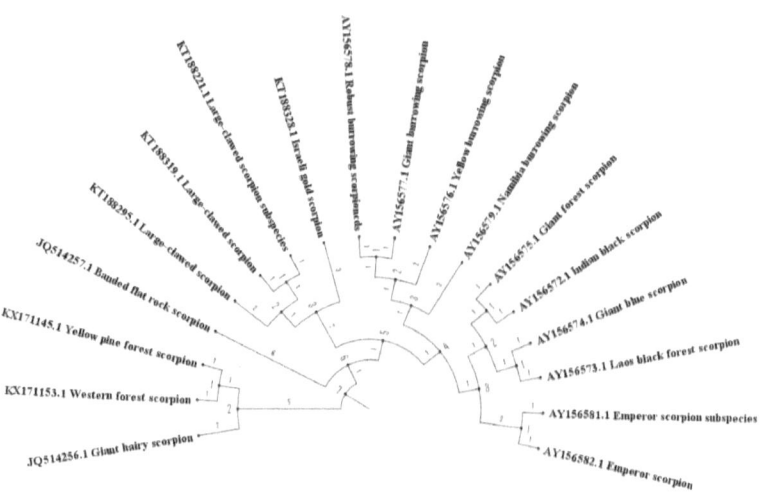

Figure 11. Phylogenetic reconstruction of 18 cytochrome
COI sequences representing order *Scorpiones*.

Discussion

A fundamental question in evolutionary biology is understanding why some groups of organisms are highly diverse. Species diversity is the result of the balance between speciation and extinction whereas morphological disparity is primarily a consequence of adaptation. [74] At the most fundamental level, mutations are essential to evolution. Every genetic feature in every organism was, initially, the result of a mutation. [75] All mutations that affect the fitness of future generations are agents of evolution. [75] Radioresitance, among other adaptive features found in organisms, can be genetically determined and inherited. [76 77 78 79]

Radioresistance is surprisingly high in many organisms. For example, the study of environment, animals and plants around the Chernobyl accident area has revealed an unexpected survival of many species, despite the high radiation levels. [80] A Brazilian study in a hill in the state of Minas Gerais which has high natural radiation levels from uranium deposits, has also shown many radioresistant insects, worms, and plants. [81] These rare examples give us a glimpse into radioresistance at a localized generational scale but say nothing about the possible influx of selective pressures that it imposes over long stretches of geological time.

We know organisms that are susceptible to radiation are often more inclined to its effects; such effects include those that cause permanent alterations of nucleotide sequences of the genome. To this end, organisms that better resist the effects of radiation also have a relative fitness advantage to its damaging effects and it too may have an impact at the population level. Selective traits are inherited. Let us take another look at *Scorpiones*, for instance.

The scorpion's carboline-composed exoskeleton is thick and durable, providing good protection from a range of environmental threats (see Images 1-4). Carboline chemicals are found in many animal cells, not just scorpions, and are thought to be sunscreens that protect epidermal cells by reflecting or scattering UV radiation. [81] Research suggests that oxygen levels were much lower when the first ancestral scorpion lineages appeared during the Silurian period; a time when more UV radiation was able to reach the Earth's surface. [82] This might explain the evolution of carboline in the exoskeletons of modern day *Scorpiones*. In addition to its carboline-composed exoskeleton, other studies have demonstrated that internal antioxidant chemistry may be responsible for shielding its cells from radiation damage. [83] What is more, scorpion morphology has changed relatively little since their first appearance in the fossil record. [84]

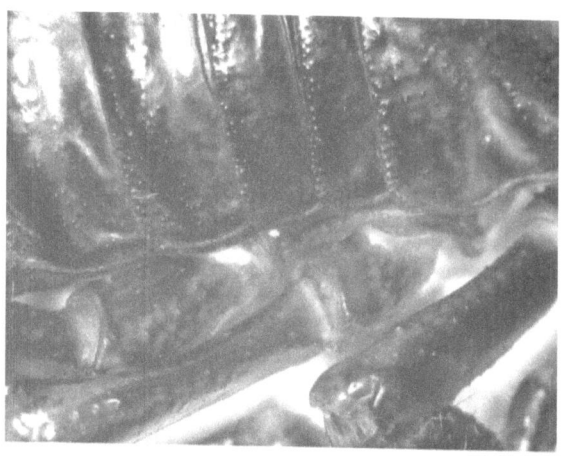

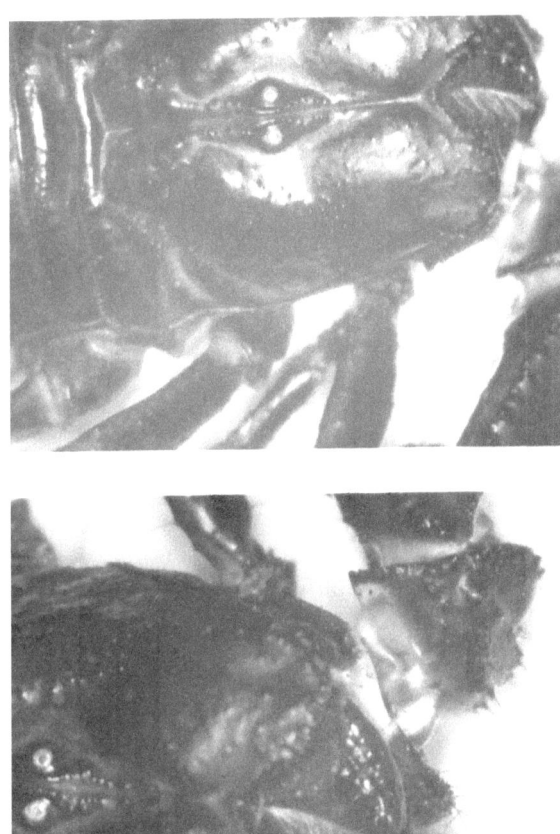

Images 1-4. Digital microscopic images of *Scorpiones* carboline exoskeleton at 30X. Species shown: *Heterometrus spinifer.*

Longwave UV light is reflected as visible light in the green range. For this reason, scorpions appear to glow in darkness. Under direct sunlight, the fluorescence may impart a greenish tint to the scorpion's color. [85] Studies have shown this feature to derive from certain molecules in one layer of the cuticle, the tough but somewhat flexible part of a scorpion's exoskeleton, which absorbs the longer wavelengths of UV light and emit it in different wavelengths that are visible at night as a blue-green glow. [86]

An experiment led by Kloock at California State University found that the fluorescence seemed to help scorpions detect and avoid UV light. [86][87] The activity level of the test subjects in this study was measured by the number of transitions from the exposed to shadowed regions, and choice was measured by the percentage of time spent in the shadowed portion of the arena under IR-UV. [87] The Kloock study interpreted their observations as evidence that fluorescence aids in the detection of and response to UV light.

Now, back to the topic of DNA. The evidence from cell science is further compelling. Consider the following: Aside from errors in cellular replication and oxidative damage, *germline* mutations can also occur due to exogenous factors that include chemical substances and ionizing radiation. [88][89][90] One well-documented study describes how single exposure or multiple exposures to a low dose radiation induced a significant cytogenetic adaptive response in mouse germ cells. [90][91] As noted above, *Scorpiones* are unique in that they combine traits for UV light avoidance along with a UV reflective exoskeleton and an internal metabolism fit for shielding radiation damage to cells. The hypersensitivity of the rodent anatomy, as with other members of the mammalian phylum, predisposes it to more of the DNA-damaging effects of naturally occurring radiation.

Biogeography is another considerable element to this study worth mentioning. One of the objectives was to demonstrate a quantifiable degree of genetic variance between two distinct groups of organisms in conjunction with, or despite of, their geographical distribution. Furthermore, it might be expected that higher degrees of variance would be present among groups of closely related organisms that occur across wider ranges in a geographical context; and the reciprocal scenario could be assumed in like manner.

Both rodents and scorpions overlap in terms of their natural environments, exposing each form to nearly equal amounts of atmospheric radiation brought on from UV light. Surprisingly, despite a much lower ratio of total pairwise dissimilarity, members of group *Scorpiones* also happen to be more biogeographically extant. Moreover, it has also been shown that members belonging to order *Rodentia*, have an increased rate of molecular evolution due to a higher mutation rate. *Rodentia* is the most diverse order of placental mammals, with extant rodent species representing about half of all placental diversity. [92]

Thus, the question now becomes whether localized radiation events contribute to an increased rate of molecular evolution in organisms that lack radioresistance. And if so, what can we infer about the opposite case scenario? Studies like these pose several unique challenges. First, the data sampling used in this study is both incomplete and relatively small to draw any significant inferences about an entire phylum of species. And, in fact, much of this interpretation could be explained in the historical context of natural phenomenon like genetic drift, retrovirus insertions, or even the vacancies left by open ecological niches; just to name a few examples. While many of these mechanisms do play invaluable roles in species diversification, this study solely ventures to assess any possible association that may exist between naturally-occurring radiation and its implications on population genetics among groups of organisms that reside on far ends of the radioresistance spectrum.

To what extent to which population genetics becomes affected by natural-occurring radiation will require further investigation. Similar studies in laboratory settings have focused on microorganisms but rarely has it included multi-cellular eukaryotes. Indeed, an argument could even be raised about the timescale limitations involved with testing multi-cellular eukaryotes, coupled with the rarity of inheritable mutation, which would make such an experiment impractical in a controlled environment. Yet, there may be a reliable framework by which to detect, assess, and quantify *radiation-induced evolution* over successive generations. Before this paper concludes, let us elaborate further on a potential experiment design.

Imagine for moment, a more ideal experiment involving three homologous sets of fast-breeding model organisms that have adjusted their sexual reproductive patterns to satisfy remarkably short life cycles. Over the course of this experiment, each group is exposed to different levels UV radiation (perhaps ranging between 180 and 400 nm); except for the control group, which remains unexposed. Such an experiment could incorporate test subjects like *Drosophila melanogaster* or *Acheta domestica*, as they are known to be good candidates for studying a wide array of biological processes. After careful match-breeding over different generations, gene sequencing is applied to each line of offspring within the same test group or population. Complete sequences of mitochondrial DNA could be a useful biomarker in this experiment, due to its tendency toward rapid mutation rates. Techniques in comparative sequence analysis can then reveal even the slightest biological change over a multitude of successive generations. In some ways and on a more simplified scale, this paper tries to recreate this hypothetical experiment without the medium of laboratory testing and procedures and only by means of design and publicly available genomic data.

Summary

Every cell in every organism is intimately connected to the environment surrounding it. Radiation occurs all around us and, depending on the amounts of energy associated with it, can cause significant biological effects observable down to a cellular scale and possibly manifested up to the population level. Over the course of Earth's history, changes to the atmosphere must have played a key role in the conditions under which life formed and evolved. [37][38] Events in the history of life do very unlikely represent a pure coincidence, but the extent to which they can be linked to the establishment of a modern atmosphere and the fluctuating levels of radiation that came thereafter is not entirely settled. [93]

While it is difficult to quantify the role of naturally-occurring radiation in shaping the outcome of living systems over the course of extended evolutionary time, the results do show a distinct correlation between higher and lower degrees of genetic variance among two different groups of organisms that reside on opposite ends of the radioresistance spectrum; where the order, group or population that has shown increased hypersensitivity to radiation also has an increased rate of molecular evolution due to a higher mutation rate. With that said, the reader should be reminded that correlation does not necessarily indicate a causation. Whether or not naturally occurring radiation is directly or indirectly implicated in creating genetic variance within populations of living organisms is subject to more research.

5

Interpreting Large-Scale Phylogenetic Models of Mammal Class Diversification Based on Mitochondrial Biomarkers

Originally published in the Journal of Bioinformatics, Proteomics, and Imaging Analysis

The great diversity of Cenozoic mammalian families includes some five thousand four hundred documented species of living mammals assigned to twenty-six orders. [94] Yet, despite the abundant amount of diversity, much remains unknown about their ancestral lineages prior to the Paleocene epoch. Several mammalian orders are only interpreted as diversifying immediately after the K–Pg boundary, including the largest animals on the planet, the great whales, as well as some of the most intelligent, such as elephants, primates, and cetaceans. [95] It is thought that modern mammals arose in the Paleogene and Neogene periods of the Cenozoic, after the extinction of non-avian dinosaurs. [96] Prior to the K–Pg boundary, mammalian species are generally found small, comparable in size to rats; this small size would have helped them find shelter in protected environments. [96]

When most of the non-avian dinosaurs perished, the

surviving mammals diversified into the dinosaurs' niches, where they remain today. [97] Interestingly, recent investigations in phylogeny research have reported two significant findings: (a) rapid diversification among primitive eutherian mammals may have occurred much earlier than traditional models presume; [98] [99] and (b) the oldest living lineages of modern eutherians can be traceable to a single modern group. [100] Several phylogenetic studies involving mitochondrial biomarkers have successfully reconstructed the phylogeny of mammals at different levels and scales of the taxonomic hierarchy to provide a basis for standardizing methodologies. [101] One such paper published in March 2016, postulates that a single group of modern mammals predates the extinction of the dinosaurs. [101] The Brandt study used two methods to sequence the traces of mitochondrial DNA collected and confirmed that it diverged from all other mammals approximately 78 Myr. Mitochondrial DNA provides the ideal test case for these kinds of analysis because all genes are inherited as a single unit and thus have a single evolutionary history. [102]

Genetic analysis has played an increasingly important role in confirming existing or establishing radically different mammalian groupings and phylogenies. [103] A combinational approach may too provide an ideal method for verification, where genetic analysis is crossed-referenced with morphological models. As it related to this chapter, in 2011 researchers reported on the discovery of a fossil mammal found in China that would have lived alongside the dinosaurs and that, at 160 million years old, represents one of the earliest mammals known today. [104] *Juramaia sinensis*, a furry rodent- like animal just a few centimeters long, is thought to be the oldest known common ancestor of modern placentals, or a very closely related cousin to that common ancestor. [104] The correlation between *Juramaia sinensis* and the findings present here will be further highlighted in later sections.

Fossils alone are not always available and sufficiently informative, and phylogenetic methods based on models of character evolution can be unsatisfactory. [105] Genomic data offer a more robust opportunity to estimate these ancestral lineages. [106] This investigation features several techniques in bioinformatics for phylogeny research. By reconstructing a large-scale phylogeny based on mitochondrial biomarkers, I seek to reaffirm models of mammal class diversification among lineages that endured through the K-T event and onward. Two hundred thirty-five complete mtDNA sequences and sixty-two major taxa families within the class mammalia were represented in this study. My primary objective is to outline a practical framework by which time-extended lineages could be assessed and evaluated, for purposes of better understanding the evolutionary trajectories that led to the abundant diversity within mammalian families.

Sixty-two taxa families for sequence selection

Mitochondrial biomarkers serve a practical use in large-scale processing for comparative sequence analysis. As such, mitochondrial sequences have the advantage of being translatable, and at the level of species and genera usually do not contain high numbers of length-variable regions. [107] Two key factors support the application of mtDNA for molecular phylogenetics: (a) as explained in chapter 2, complete mtDNA datasets are light-weight compared to larger genomic datasets that can create systematic bottlenecks during processing and execution that lead to erroneous inferences; and (b) mitochondrial DNA accumulates nucleotide substitutions relatively rapidly, due to lack of repair mechanisms that slow down the molecular clock. [108] In practice, this feature makes complete mtDNA sequences suitable for inquiries involving species-level and genus-level classifications.

My mtDNA sequence selections relied exclusively on preexisting order classifications, as to represent the major family groups within the class mammalia. The NCBI nucleotide databank was the platform where each mtDNA sequence was collected.

On several different instances, BLAST similarity searches were required to identify the most homologous sequence candidates among the many mammalian groups selected for purposes of this study. And, species selection was appropriated based on cross-referencing independent data with respect to morphology alone. From these findings, several distinct FASTA data files containing a combination of two hundred thirty-five mtDNA sequences that represent the sixty-two major families were compiled. Each individual data file was later combined into a master file that contained all the family types outlined later this chapter. It should then be noted, neither file exceeded 4,000 KB.

A large-scale phylogenetic tree

On a broad spectrum, the phylogenetic trees generated by the UGENE software depicts an evolutionary scenario consistent with hierarchal models of mammal class diversifcation; and it highlights the divergent events that further distinguish the three modern groupings within the class mammalia: 1) monotremes; 2) marsupials; and 3) placentals. Sister taxa among clade distributions are arranged in accordance with morphological classifications. Each taxonomic unit with descendants also constitutes the inferred most recent common ancestor of the descendants and the edge lengths may be interpreted as rough time estimates.

Located on the internal branches near the root, we find nodes that represent the oldest common ancestor of Cretaceous mammals, presumably. As we move from one internal node to the next, the variation of genetic distance increases minutely and only widens at the genera level. Similarity ratios show a marginal disparity – ranging between + - 65 to + - 92 percent – among seven family types concentrated nearest the root; and it shows a + - 50 to + - 64 percent divergence from taxonomic units within the remaining clades. The identity distance matrices shown on Figure 12 better illustrate the measurements of genetic divergence between each sequence, where the final distance value is the average of PHYLIP neighbor-joining estimates. In the context of extinction events and the ecological niches that are filled thereafter, we might expect to observe widening gaps in degrees of genetic variation following a period of rapid diversification. This chapter demonstrates it on several different scales.

Maximal node measurements further support that interpretation. Estimates of divergence are not always straightforward and the rate of evolution is not uniform in different lineages. [109] Nonetheless, it is widely held that trees exerting node measurements of 0.75 or higher are generally reliable, in terms of Bayesian-obtained approximates. [110] A high value means that there is strong evidence that the sequences to the right of the node cluster together to the exclusion of any other. [111] Arguably so, measurements that do not meet a minimal requirement for posterior probabilities in maximal node measurements can lead to erroneous inferences. [112] [113] [114] In this case, the maximal measurement values show a steady rate of calibration consistency, represented by the maximal support measurement per internal node illustrated on Figure 14. Each maximal support measurement is 1.0, regardless of node placement within the tree. As such, the resulting cladogram shows localized diversification events in multiple time periods, across multiple scales.

Exact time estimates, however, cannot be determined by these figures alone. The molecular clock, shown here by the internal branch lengths and their respective node ages, suggests that placental lineages are far older than traditional models presume; and their diversification may be linked prior to the breakup of the continents before the end of the Cretaceous period. [98] At any rate, the genomic data needed to narrow the scope further was not readily available to this investigation.

Figure 12. Complete Phylogenetic Tree of Sixty-Two Distinct
Mammalian Families Based on Complete mtDNA Sequences
(17,240 bp). Includes: *Aotidae, Atelidae, Balaenidae, Balaenopteridae,
Bathyergidae, Bradypodidae, Bovidae, Caenolestidae, Callitrichidae,
Camelidae, Canidae, Castoridae, Cebidae, Cercopithecidae, Cervidae,
Chinchillidae, Chlamyphoridae, Cricetidae, Dasypodidae, Delphinidae,
Didelphidae, Dugongidae, Elephantidae, Equidae, Eupleridae, Felidae,
Giraffidae, Herpestidae, Hippopotamidae, Hominidae, Hyaenidae,
Indriidae, Iniidae, Kogiidae, Lemuridae, Leporidae, Lipotidae,
Macropodidae, Megalonychidae, Muridae, Mustelidae, Myrmecophagidae,
Ornithorhynchidae, Orycteropodidae, Otariidae, Phalangeridae,
Phascolarctidae, Phocidae, Pitheciidae, Procyonidae, Physeteridae,
Pteropodidae, Rhinocerotidae, Rhinolophidae, Sciuridae, Soricidae,
Suidae, Talpidae, Tapiridae, Trichechidae, Ursidae, Vespertilionidae.*

Discussion

The "explosive model" of mammal evolution proposes that late placental lineages emerged and diversified to fill niches left vacant after the KT catastrophe. [115] Conversely, the cladogram illustrated above (Figure 12) depicts a rather different trajectory. My results suggest that placental lineages are possibly older than traditionally presumed, hinting their diversification was linked prior to the breakup of the continents before the end of the Cretaceous period. [116] Although that interpretation is not new to the field of phylogenetics, it is an independent confirmation of the inferences raised by others that reinforce the early divergence of mammals. [117] [118]

As shown on Figure 12, seventeen distinct organisms belonging to four family types represent the closest living lineages from which placental mammals and marsupials diverge from a common ancestor: *Cricetidae, Muridae, Soricidae,* and *Talpidae.* This group of organisms is interesting because of its retention of primitive traits, like *Juramaia sinensis*; the oldest eutherian fossil species ever found. [119] [120] Lines of anatomical evidence support the idea that *Juramaia sinensis* is closely related to placental mammals. [120] For example, *Juramaia sinensis* has three molars and five pre-molars — like placentals, but unlike marsupials which have four molars and three pre-molars. [120] Before the discovery of *Juramaia sinensis*, the earliest known fossil relatives of placentals dated to around 125 Myr. [119] [120] Since each lineage clearly existed as a distinct entity approximately 125 Myr, the divergence between placentals and marsupials may have occurred sometime before then, presumably. [120]

My findings are not necessarily indicative of a direct living lineage to a common ancestor of a placental in the form a single-family type but could also be explained by missing ancestral unit(s) not present on the cladogram. Other phylogenies, most notably those reconstructed on morphology, often disagree with respect to time estimates and divergence in mammal evolution. But unlike other models that suggest explosive evolution in post-Cretaceous context, the scenario outlined by my results predict a series of rapid diversification events in multiple time periods, and should therefore not be confused with a single "explosive" radiation event preceding or following the K–Pg boundary.

Lastly, something should be said about the sequence selection in phylogeny research. A handful of studies have well documented the limitations of using mtDNA to reconstruct phylogenies that involve time-extended lineages. Rapid mutation rates in mtDNA produce significant molecular variance among immediate populations. [121] This has notable benefits when studying ancestral relationships whose divergence times are thought to be no greater than 8 to 10 Myr. [122] However, in reciprocated cases, efficacy may become less resolved beyond that scope. In addition to that, hybridization effects can cause mtDNA to move freely between species, which could infer incorrect relationships when building phylogenies. [123]

Despite these legitimate critiques, this paper addressed the practical applications of using mitochondrial datasets in large-scale studies. I would also point out that examples of natural hybridization leading to speciation are exceedingly rare, as mentioned in earlier chapters. As mentioned in chapter 2, while most known cases of hybrid speciation occur in plants, most documented cases involving animals have been observed in fish and insects. [124] Perhaps most important in this discussion, mtDNA datasets have repeatedly shown to provide enough sufficient resolution for reconstructing a robust phylogeny and it also facilitates the molecular dating of divergence events within a phylogeny. [125] Mitochondrial DNA is particularly useful in phylogenetic studies, as it demonstrates high interspecies conservation and at the same time is variable enough to allow intraspecies differentiation. [126]

Summary

By assessing the number of node measurements, where the taxonomic units with higher nucleotide substitution rates reside on the far ends of internal branches, and by interpreting the molecular clock in accordance with their respective node ages, I find very good support for my original inquiry. Based on this interpretation, the resulting cladogram provides evidence for localized diversification events in multiple time periods, across multiple scales. This inference also qualifies the second part of this study, which holds that the closest living lineages of early placentals derive from a specific group that shares homologous traits with the oldest eutherian fossil species ever found. The lack of data needed to narrow the scope further makes an accurate time estimate difficult to acquire; but these results may provide a reliable starting point for a more comprehensive investigation.

6

A Phylogenetic Examination of Non-Native Freshwater Fish Populations in Florida

Written for Invasion Ecology of Aquatic
Animals, University of Florida

In the field of invasion ecology, there remains significant gaps in our understanding of the prerequisite set of characteristics that cause a non-native or exotic species more likely to become established in a new environment versus another that fails in doing so. Several factors have been identified as potentially playing a major role. Lack of predation, competition, disease, reproductive rates, increased lifespan, dispersal ability, open niche exploitation, biotic resistance, species richness, and a generalist's lifestyle are some of the variables that are thought to be most important in the invasion process. [127] [128] [129] Additionally, something should be said about preexisting similarities in climatic conditions between the introduced species' region of origin and its new geographical habitat. Geographical proximity paired alongside human activities have also been implicated in contributing to subsequent establishments.

This paper specifically addresses these matters within the confines of freshwater fish populations in Florida. Florida is well known to be a breeding ground for many non-native freshwater fish varieties that, in some cases, even reach the level of invasiveness. The focus here will be to examine various degrees of phylogenetic relatedness among native and non-native freshwater fish species and its role as a predictive indicator for successful establishments. Genomic sequences are the primary biomarkers for building a phylogenetic tree as a reference point for this exercise. In doing so, this data-driven investigation attempts to answer the following inquiry: Do higher or lesser degrees of phylogenetic relatedness between native and non-native species contribute to a reliable risk assessment prediction?

Other studies have looked at these possibilities. Papers that regard genetic variation as an indicator for the success of an invasive species have produced mixed results. It was Darwin's naturalization hypothesis which first predicted that the less closely related native species are, the more likely they are to succeed as invaders. [130] To be clear, however, Darwin was referring to species of flora; not alike the aquatic fauna this paper will examine. It is also worth noting that Darwin had no prior knowledge of genetics. The science was simply not available at the time.

Despite being uniformed about the role of genetics, several modern studies support Darwin's hypotheses. For example, a recent paper by the Carnegie Institution of Washington found that exotic taxa less related to native species are more invasive. [131] The Strauss study began by asking whether already established introduced species with lesser impacts differ in their phylogenetic relatedness to native species than do high-impact, invasive species. Other researchers were also surprised to find ecological success among their model organisms given that reductions in genetic diversity are generally believed to be harmful. [132] In summary, these matters can be further examined utilizing a comparative methodology. By taking a phylogenetic approach to native and non-native freshwater fish populations in Florida, this chapter addresses whether phylogenetic relatedness is a relatively reliable predictor of the spread and establishment of non-native or exotic species.

Sequence selection

In terms of assembling a dataset, species selection comprised of native and non-native aquatic species that share climatic similarities, geographical proximity, taxonomic classification, and relative size and weight. For purposes of also including a control group, genomic sequences were broken into three distinct categories based on their current biogeographical statuses in Florida: (1) native species; (2) non-native species, established; and (3) non-native species, introduced with limited distribution and subsequent unsuccessful establishments. *Cichlidae* is the most represented group of nonindigenous freshwater fishes in Florida, and therefore, composing of the largest non-native family in the entire dataset. Native species were mostly selected from the family of *Centrarchidae*, due to their native dominance in respective numbers. Other varieties include species from the family of *Serrasalmidae* and *Cyprinodontidae*, which overlap collectively in terms of their biogeography among the other selected candidates. A total of twenty-two sequences were used in this study.

Again, with respect to genome type, mitochondrial DNA (mtDNA) serves as a suitable biomarker for this study due to its elevated mutation rate, lack of recombination, and maternal inheritance attributes. [133] As noted numerous times, when building a phylogenetic tree from raw genomic data, mtDNA is often considered a practical genome candidate for this type of exercise. DNA sequences from the mitochondrially encoded genes are attractive sources of characters for estimating the phylogenies of recently evolved taxa because mtDNA evolves rapidly. [134] In this study, each mtDNA sequence was collected and assembled into a consolidated dataset via the NCBI Nucleotide Database search. Moreover, the mtDNA sequences used in this investigation were compiled in FASTA format for Multiple Sequence Alignment (MSA). Table 5 shows species selection, current biogeographical status, and sequence accession number.

Common Name	Scientific Name	Native	Non-Native Established	Non-Native Not Established	Accession Number
Aureum Cichlid	*Thorichthys aureus*			✓	NC031182.1
Bay Snook	*Petenia splendida*			✓	KJ914665.1
Black Crappie	*Poxoxis nigromaculatus*	✓			MH324430.1
Blue Discuss	*Symphysodon haraldi*			✓	KT215609.1
Blue Eyed Cichlid	*Cryptoheros cutteri*			✓	KR150878.1
Bluegill	*Lepomis macrochirus*	✓			MF621714.1
Butterfly Peacock Bass	*Cichla ocellaris*		✓		KR150863.1
Florida Flagfish	*Jordanella floridae*	✓			AP006778.1
Florida Largemouth Bass	*Micropterus floridanus*	✓			MH301073.1
Jack Dempsey	*Rocio octofasciata*			✓	KR150870.1
Krobia	*Krobia guianensis*			✓	KR233978.1
Midas Cichlid	*Amphilophus citrinellus*		✓		KJ562277.1
Oscar	*Astronotus ocellatus*		✓		NC009058.1

Pearl Cichlid	*Geophagus brasiliensis*				✓	NC031181.1
Ram Cichlid	*Mikrogeophagus ramirezi*				✓	KR233976.1
Redbreast Sunfish	*Lepomis auritus*	✓				MH301066.1
Redhump Eartheater	*Geophagus steindachneri*	✓				KR150866.1
Small Scaled Pacu	*Piaractus mesopotamicus*				✓	NC024940.1
Spotted Sunfish	*Lepomis punctatus*	✓				MH301069.1
Texas Cichlid	*Herichthys cyanoguttatus*				✓	NC033546.1
Triangle Cichlid	*Uaru amphiacanthoides*				✓	KR150875.1
Warmouth	*Chaenobryttus gulosus*	✓				NC042249.1

Table 5. A dataset comprising of twenty-two native and non-native Florida freshwater fish species. Taxonomic families include *Centrarchidae*, *Cichlidae*, *Serrasalmidae*, and *Cyprinodontidae*. Current non-native population statuses were obtained from a list of exotic freshwater fishes in Florida. [135]

Results

Phylogenetic analysis usually requires that we begin with a null hypothesis. As mentioned at the beginning of this chapter, similar studies have produced mixed results. The cladogram seen on Figure 13 illustrates a highly probable distribution of species in terms of their evolutionary history. For instance, native varieties belonging to a specific family, namely *Centrarchidae*, are grouped together within their own respective clade nearest a sister clade containing a native species of *Cyprinodontidae*.

Non-native *Cichlidae* outnumber every additional family type in this assortment. Though, only three of fourteen cichlid species meet the experiment criteria for both non-native and established populations (see Table 5). These species include: (1) *Cichla ocellaris*; (2) *Amphilophus citrinellus*; and (3) *Astronotus ocellatus*. Interestingly, the nodes that represent these three populations are found in placements along the far outgroups of the tree within the larger scope of the *Cichlidae* family clade. *Astronotus ocellatus* being the most phylogenetically distant of the three species, followed closely by *Cichla ocellaris*. *Amphilophus citrinellus* is nested deeply within the *Cichlidae* family clade on the opposite end of the cladogram. To put it simply, the arrangement of nested lineages in this cladogram show that three established non-native populations of cichlids trend in the direction of being more phylogenetically distant than other non-native (not established) species located closer to the inner clades containing native taxa. The branch length values associated with each outgroup lineage is also another indicator of a deeper ancestral divergence event. Thus, based on these results, increasingly higher degrees of genetic variation (also interpreted as phylogenetic relatedness in this context) are found to correlate with the successful establishment of three distinct non-native freshwater fish species in Florida.

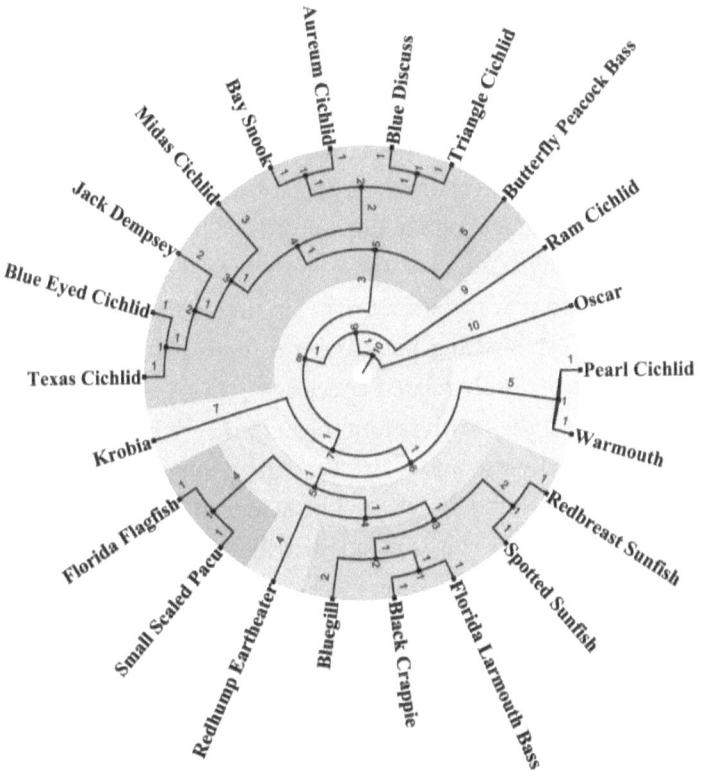

Figure 13. Resulting cladogram composed of twenty-two native, non-native, or introduced freshwater fish species in Florida.

Discussion

All *Cichlidae* share common ancestry to populations that first appeared when South America and Africa were geographically connected. [136] After the split of the two land masses, cichlid populations on both continents diversified into hundreds of different species within their respective tropical ranges. Adaptations to warmer waters, lowly dissolved oxygen, and desiccation make certain varieties of cichlids ideal candidates for Florida habitats. [128] [137] Many cichlids share and display characteristics that might define them as fierce competitors. Unlike some native freshwater fish species in Florida, cichlids also engage in some form of parental care for their offspring, either mouthbrooding or otherwise defending eggs and larvae. [128] These homologous traits have been observed and documented in their native ranges, newly introduced habitats, and controlled laboratory settings. Yet, even these shared set of characteristics do not fully explain why some species belonging to the same *Cichlidae* family fail in establishing permanent foreign populations versus other successful closely related species.

Even among their own taxonomical ranks, cichlids themselves are genetically diverse and share very little in common with the *Centrarchidae* of North America. The extent of these relationships is well-illustrated in the isolated distribution of clades seen in the cladogram labeled Figure 13. Despite any deep preexisting evolutionary events that led to their modern biogeographical distributions, these highly diverse and evolutionarily distant families now share a common Floridian environment. Genetic variation is often cited in studies for its involvement in the success of healthy populations of organisms. Aside from variation, it is thought that phylogenetically related species share similar ecological functions. [138] Going back to Darwin's naturalization hypothesis, the suggestion that novel genera would be more successful in naturalizing in new ranges than genera with native representatives has been supported by studies that focus on applications involving comparative sequence analysis. [131] Because a large diversity of other mechanisms may also underlie establishment success, ecological novelty could be partially gauged by the phylogenetic relatedness between native inhabitants and non-native invaders and is a general metric that does not require knowledge of specific traits needed for adaption. [131]

More supporting evidence is found in studies that examine phylogenetic patterns across different spatial scales, ecosystem type, and taxa. [138] Not unlike the results described in this paper, a 2016 showed that exotic species more closely related to natives tended to be less successful; their analysis consistently generated negative relationships between invader–native phylogenetic relatedness and invader success at the local scale, but not the regional scale. [138] While this study looked at spatial implications combined with phylogenetic analysis, a 2011 Stellenbosch University investigation of California and Florida ecosystems concluded that successfully introduced species are more distantly related to natives than failed species, although variation was high. [139]

Summary

In conclusion, the compilation of independent studies, including others not mentioned here, improves our understanding of biological invasions by examining the benefits from the use of community phylogenetic approaches. [140] The application of phylogenetics toward invasion ecology could provide a useful criterion for prioritizing management efforts toward the introduction of non-native species. [131] On an important side note, this study only encompassed a small sample-sized group of aquatic organisms. Significant taxonomical gaps among the freshwater fish species in this dataset could have implications on the results. This might explain a couple irregularities found among overlapping lineages in the cladogram above. Let us not forget that these modern groups, taxa, or families have been separated by geographical barriers that span millions of years of biological evolution. Nonetheless, given a small dataset and limited timeframes to expand upon it further, these findings support a positive correlation between varying degrees of phylogenetic relatedness and the spread and establishment of non-native species in Florida. To be fully confident about such inferences, an in-depth investigation is recommended utilizing a largely assembled dataset containing a wider scope of species, taxa, and genera.

7

PHYLOGENETIC ANALYSIS OF MATERNAL LINEAGES IN MODERN-DAY BREEDS OF BRITISH CANIS LUPUS FAMILIARIS

Originally published in the International Journal of Research Studies in Biosciences

The origin of domesticated *Canis lupus familiaris* is not clear. Its closest living relative is the gray wolf (*Canis lupus*), and there is no evidence of any other canine contributing to its genetic lineage. [141] [142] [143] Studies propose a divergence time of the dog from the wolf ancestor at 27,000 YA, with the most recent estimate of domestication occurring between 20,000 and 40,000 YA. [144] The cohabitation of dogs and humans would have greatly improved the chances of survival for early human groups, and the domestication of dogs may have been one of the key forces that led to human success. [145] The oldest dog breeds evolved or were bred to fill certain roles. [146] A recent paper published in the journal *Science* said domestication likely happened from two separate wolf populations, one in Europe and the other in Asia. [147] British breeds are of a special interest, due to their many well-established varieties.

A robust phylogeny of post-Victorian era British dog breeds

has not yet been established. Other phylogenies have looked at the relationship between modern-day dog breeds on broad scales, including European varieties, but neglect to detail geographically isolated populations. This chapter focuses primarily on comparative techniques in phylogeny research to best infer the earliest members of, or most closely related modern-day breeds to the original varieties on the British Isles. Here, I examine maternal lineages among post-Victorian dog breeds by way of mitochondrial DNA (mtDNA) sequences. Two distinct sets of raw sequences were used in this investigation; one consists of complete genomes, whereas the other set is made up of partial sequences. My results will show that modern-day hounds and herding breeds fall closest to the midpoint of unrooted trees, making these individual varieties the closest living relatives to the oldest living lineages of European ancestry in Britain.

Sequence Selection

As with every study cited in this book, the NCBI nucleotide databank is the repository where each mtDNA sequence was acquired. BLAST similarity searches regularly facilitated identifying homologous sequence candidates among closely related groups. Each set of raw sequences are referenced in 4 studies: (1) Sequence Diversity of the Canine Mitochondrial Genome; (2) Mitochondrial genome DNA analysis of the domestic dog: identifying informative SNPs outside of the control region; (3) Identification of Single Nucleotide Polymorphisms within the mtDNA Genome of the Domestic Dog to Discriminate Individuals with Common HVI Haplotypes; and (4) Forensic Informativity of ~ 3000 bp of Coding Sequence of Domestic Dog mtDNA. [148] [149] [150] [151] From these collective findings, 2 distinct FASTA files containing a combination of 36 mtDNA sequences were compiled. Among the 36 mtDNA sequences, 15 sequences were available in complete format, while the remaining sequences were collected in partial format. See Table 6 for references, annotation numbers, and sequence descriptions.

Unrooted Trees for Phylogenetic Analysis

Domesticated dogs are byproducts of controlled breeding practices that make it difficult to establish a historically accurate geographical point of origin, even among the oldest lineages on a phylogenetic tree. Some statistical evidence suggests that Asian and Middle Eastern varieties diverged long before what has been called the 'Victorian Explosion' of dog breeds in Britain and the rest of Europe. [152] One such study identified 9 dog breeds that could be represented on the outgroup of a broad-scale inter-species phylogeny. [153] The Pollinger study examined 48,000 single nucleotide polymorphisms that gave a genome-wide coverage of 912 dogs representing 85 breeds. Among the oldest lineages of European descent were those most closely related to herding breeds, mastiffs, and other hound varieties. [152][153]

Up until now, the quality of a detailed geographical analysis is sparse and unclear. Examining a localized dog population within continental Europe presents a challenge to a phylogenetic assessment, due to crossbreeding occurrences beyond regional boundaries. Such phylogenies are impractical because they do not resolve any evolutionary derived clades, because of artificial selection practices. What is more, it is unlikely that a single source in the form of a common ancestor could be identified among European breeds, beyond *Canis lupus*. When a single ancestral root cannot be inferred nor assumed, a traditional cladogram does not provide the most ideal resolution. Instead, we could illustrate the relatedness of the leaf nodes without making assumptions about deep ancestry.

Britain's unique geographical features make it a particularly interesting candidate for this type of phylogenetic analysis. The coastline and landscape of what would become modern-day

Britain began to emerge at the end of the last Ice Age around 10,000 YA. [154] What had been a cold, dry tundra on the north-western edge of Europe grew warmer and wetter as the ice caps melted. [154] It was not until 6100 BC that Britain broke free of mainland Europe, during the Mesolithic period. [154] The earliest evidence of human presence in Britain is dated to 10,500 BC. [155] [156] By around 4000 BC, the island(s) became populated by people with a Neolithic culture that had already been engaging in dog domestication. [157] The oldest British dog remains were found at Star Carr Yorkshire and are dated at 7538 BC. [158]

We might then imagine a scenario where the first domesticated dogs of Britain experienced a form of reproductive isolation from a secondary population on the mainland; this was initial and not permanent. Over time, isolation can exert unique evolutionary forces that result in the development of a distinct genetic reservoir. [160] Geographical barriers would not prevent the later influxes of foreign cultures to the British Isles, including the Romans, Anglo-Saxons, and Vikings, just to name a few. Late arrivals introduced foreign dog breeds to the native population; and this too has been well-documented by historical archives. [161] [162] Alas, we now have dog varieties that are crossbreeds between one and many distinct others, and not direct decedents of one primary lineage of the original inhabiting population.

This brings me back to my position regarding the best approach for illustrating a phylogeny. As it happens to be in this case, an unrooted tree best fit the profile for a phylogenetic assessment. This method can be generated from rooted trees by simply omitting the root. [163] An unrooted tree should give no information about the order of speciation events. [164] Because the scope of this investigation does not necessarily rely on inferences regarding evolutionary history through divergent events, I look instead for patterns of relatedness among clustered networks to identify the sequence(s) with the highest nucleotide substitution rates. In terms of unrooted trees, this would be represented by the leaf node(s) most closely located to a midpoint on the diagram.

The unrooted tree(s) based on PHYLIP neighbor-joining polynomial values (Figure 14 & Figure 15) illustrates a clear divide between different groups that reside on each end of the midpoint. My results found a distinct phylogenetic clustering within modern-day British dogs that largely correspond to phenotype or function; with two exceptions. Most terriers are grouped with other terriers, hounds with hounds, spaniels with spaniels, and herding breeds reside together on both sides of the midpoint. Collectively, these phylogenies are divided into two overlapping networks of breed types: (1) terriers, mastiffs, and setters; and (2) hounds, herding breeds, and spaniels. As Figure 14 & Figure 15 show, 17 out of 19 clades could be correctly assigned to their breed based on their genotype alone. The node(s) most closely located to the midpoint on Figure 14 & Figure 15 belong to the group of hounds, including the beagle, greyhound, whippet, and the Scottish deerhound, respectively. It may be worth noting that early historical accounts describe native British dogs that are hound-like in appearance. [165] For instance, small hounds are mentioned in the Forest Laws of Canute. [165] If genuine, these laws would confirm that beagle-type dogs were present in England prior to 1016, but it is likely the laws were written in the Middle Ages to give a sense of antiquity and tradition to Forest Law. [165]

Spaniels reside nearest the midpoint in proximity to hounds, but on the adjacent end of the diagram (Figure 15). Theories on the origin of British spaniels span further back, as it is believed that Welsh spaniels are direct descendants of the *Agassian hunting dog* described in the hunting poem Cynegetica attributed to Oppian of Apamea, which belonged to the Celtic tribes of Roman-occupied Britain. [166] Sheepdogs (or herding breeds) are grouped collectively alongside hounds and spaniels within a sister taxa cluster on Figure 15, whereas terriers and setters make up the entire network on the other end of the unrooted tree. All herding behavior is modified predatory behavior; and humans began domesticating herding/working dogs during the Neolithic period. [167][168] Lastly, I should point out that Figure 15 is comprised of partial sequences. Furthermore, erroneous node placements cannot be ruled out entirely. Branch lengths and molecular clock estimates are also inapplicable due to phylogeny diagram selection.

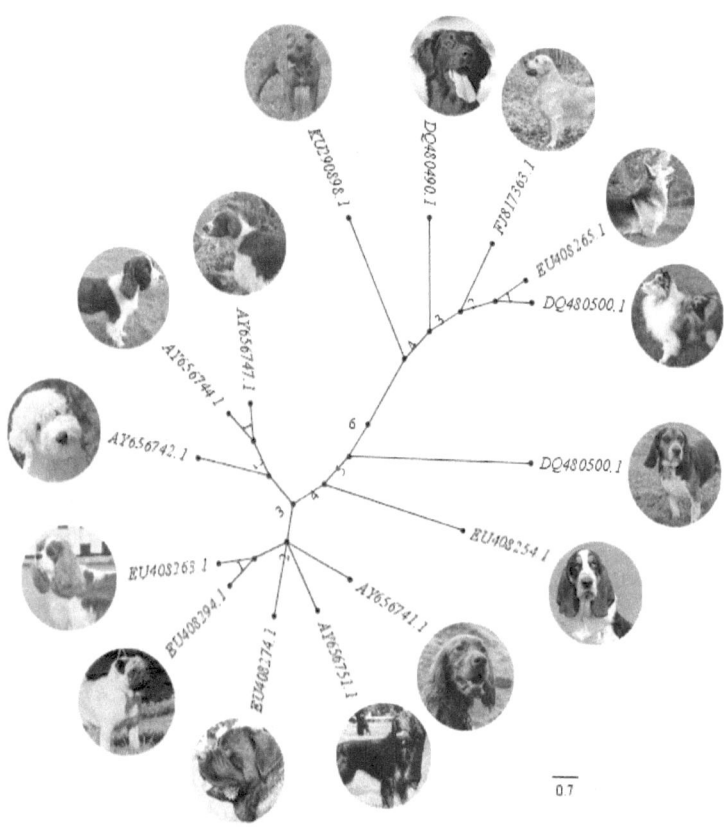

Figure 15. Unrooted trees of complete mtDNA sequences.

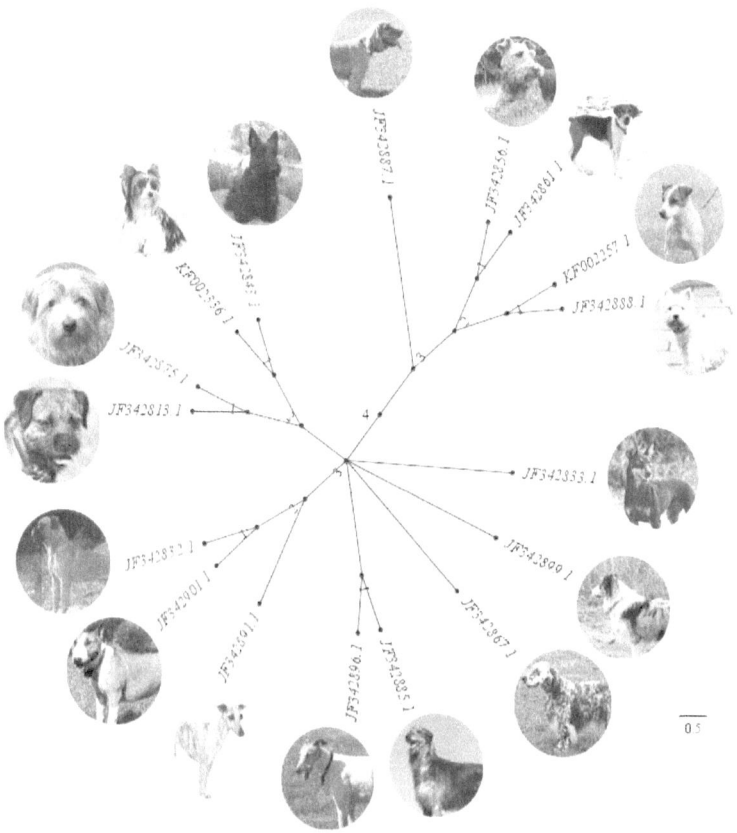

Figure 16. Unrooted trees of partial mtDNA sequences.

Pairwise Identity Ratios

Homologous sequences have identities, and degrees of conservation and similarities are quantitative. Thus, by combining the alignments and comparing each sequence against *Canis lupis* mitochondrion [complete genome], we have a quantitative reference in which to better understand the variation behind these unrooted trees. With respect to similarity ratios, Table 6 demonstrates a present correlation among the overlapping networks represented on the diagram(s) above; where the beagle (99%), Old English sheepdog (99%), Shetland sheepdog (99%), and Welsh spaniel (99%) contain the highest nucleotide substitution rates in the entire alignment.

Annotation Number [breed]	Complete Sequence	Partial Sequence
KF857179.1 [gray wolf]	-	-
AY729880.1 [beagle]	0.99	-
EU408265.1 [Welsh Corgi Pembroke]	0.98	-
AY656742.1 [Old English sheepdog]	0.99	-
JF342896.1 [greyhound]	-	0.98
DQ480500.1 [Shetland sheepdog]	0.99	-
JF342891.1 [whippet]	-	0.98
JF342885.1 [Scottish deerhound]	-	0.98
AY656747.1 [Welsh Springer spaniel]	0.99	-
AY656744.1 [English Springer spaniel]	0.99	-
EU408274.1 [English mastiff]	0.98	-
FJ817363.1 [golden retriever]	0.98	-
KU291081.1 [Labrador retriever]	0.98	-
JF342887.1 [English cocker spaniel]	-	0.97

JF342867.1 [English setter]	–	0.98
EU408254.1 [basset hound]	0.98	–
JF342861.1 [Harrier]	–	0.97
JF342856.1 [Welsh terrier]	–	0.97
JF342888.1 [West Highland white terrier]	–	0.97
JF342843.1 [Scottish terrier]	–	0.98
JF342875.1 [Norfolk terrier]	–	0.98
JF342833.1 [Manchester Terrier]	–	0.98
JF342899.1 [border collie]	0.98	–
AY656751.1 [Gordon setter]	0.98	–
AY656741.1 [Irish setter]	0.98	–
EU408294.1 [pug]	0.97	–
EU408263.1 [Cavalier King Charles spaniel]	0.98	–
KU290898.1 [Staffordshire bull terrier]	0.98	–
JF342901.1 [bull terrier]	–	0.98
JF342832.1 [bullmastiff]	–	0.98

Table 6. Results of pairwise sequence alignment.

Summary

Natural selection and selective breeding reinforce certain characteristics in dog populations, giving rise to dog types and dog breeds. In any such case, crossbreeding might make it difficult to establish a historically accurate phylogeny of divergent events among any localized group. In Britain alone, a recurring theme of outside cultural influxes may have significantly altered the evolutionary trajectory of the native breed population at any given period. Despite the difficulties of arriving at an accurate inference, this study found distinct phylogenetic clusters within modern-day British dogs that largely corresponded to phenotype or function. The unrooted tree(s) which are based on PHYLIP neighbor-joining polynomial values illustrates a divide between two overlapping networks. Falling nearest the midpoint, hounds, herding breeds, and spaniels represent a strong possible candidate for oldest living lineages in Britain; with modern-day beagles, sheepdogs, and Welsh spaniels having the highest degree of pairwise similarity ratios compared against *Canis lupis*.

8

A Review on the Evolutionary Trajectories of mRNA BRCA1/2 Genes in Primates and the Implications of Cancer Susceptibility Variants within Immediate Human Populations

Originally published in the Journal of Applied Life Sciences International

Heredity accounts for 5 to 10 percent of all cancer-related cases. [169] Some individuals are born with an increased risk of cancer because they inherit an altered gene important for cell growth or for repair of damaged DNA. [170] Over the past couples of decades, researchers have identified several gene alterations that predispose those individuals to various cancer types. [171] A well-documented example are the BRCA1/2 mutations that increase the risk of breast, ovarian and other cancers in those that inherit an altered gene. [172 173 174 175]

Most cancers do not occur in patients with a hereditary risk, but inheritance could make the risk many times higher for developing breast or ovarian cancer during the course of a lifetime. [172] Researchers have also shown that cancer susceptibility in BRCA1/2 occurs at disproportional rates across a wider range of species, particularly non-human primates. [176] In one such study, Puente revealed that human cancer alterations present in non-human primates, contain intact similar open reading frames, and show a high degree of conservation between closely related species. [176] However, it was also shown that the incidence of cancer in non-human primates is very low compared to humans. [176] While the prevailing view remains that [breast & ovarian] cancer is best and most often attributed to the damaging effects of epigenetic mechanisms that lead to abnormal cell growth, one might make a case for correlating perspectives in terms of evolutionary history. [177] Positive selection has been previously cited for its involvement in rapidly evolving regions of the BRCA1/2 variants in primate populations. [178] [179] [180]

This investigation incorporates two primary scopes for examining BRCA 1/2 cancer susceptibility among closely related lineages: (1) phylogenetic reconstruction of mRNA BRCA1/2 genes (both cancer susceptible and non-cancer susceptible) in variously distinct primate families (including *Homo sapiens*); and (2) pairwise comparative analysis of breast cancer 1 early onset BRCA1 mRNA partial cds within immediate human populations. By evaluating the results obtained from pairwise identity ratios, consensus comparisons, and phylogenetic reconstruction of 10 primate families, this chapter examines whether altered BRCA genes (or the specific mutation types) that disrupt error-free DNA repair mechanisms occur more frequently beyond interspecies taxonomy or originate spontaneously within a localized gene pool.

Sequence Selection

Two independent data types were used in this study. Each set of raw genomic data pertain to a specific experiment design. Sequence selection toward phylogenetic reconstruction compares non-cancer associated sequences against cancer susceptibility variants of the same mRNA type, among 10 primate families. As described above, tumor suppressing BRCA1/2 genomic datasets were selected due to their involvement in several different cancers in human populations and closely related others. Moreover, breast cancer 1 early onset BRCA1 mRNA variants were selected for consensus comparisons among *Homo sapiens* sequences.

Each set of raw sequences are referenced in 5 primary studies: (1) Emerging roles of BRCA1 alternative splicing; (2) Evidence of a warfarin-sensitive cancer procoagulant in V2 carcinoma; (3) Rapid evolution of BRCA1 and BRCA2 in humans and other primates; (4) Growth retardation and tumour inhibition by BRCA1; and (5) Association of BRCA1/2 mutations with ovarian cancer prognosis: An updated meta-analysis. [177 181 182 183 184] From these collective findings, 3 distinct FASTA files containing a combination of 36 genomic sequences were compiled. Breast cancer 1 early onset BRCA1 mRNA variants were obtained via NCBI nucleotide databank and were appropriated toward consensus/dissimilarity statistics.

It should be noted that the University of Utah, BRCA Mutation Database was an important reference repository in this study. See Table 7 and Table 8 for references, annotation numbers, and sequence descriptions.

Germline Mutations & Evolutionary Trees

Genetic mutation is the raw material needed for biological evolution to occur. The amount of time during which mutations accumulate to generate diversity results in higher or lesser degrees of genetic variation between different populations. It is widely held that most mutations play no significant role in the evolutionary process. Only those mutations that occur to the germline are significant in terms of evolutionary change. We assume that most mutations – something in range of 90 percent or higher – are either neutral or harmful to a host organism. [185 186 187]

Germline mutations in the BRCA1/2 genes predispose affected individuals to breast and ovarian cancer syndromes. [175] The National Cancer Institute cites three reported founder mutations in BRCA1/2: (1) BRCA1:c.68_69delAG, (2) BRCA1:c.5266dupC; and (3) BRCA3:c.5946delT. [188] Both BRCA1 mutations are known founders in the Ashkenazi Jewish population, with c.68_69delAG being the most frequent with approximately 0.9% of all Ashkenazi Jewish individuals being carriers. [188] BRCA1:c.68_69delAG is found most frequently in individuals of Ashkenazi Jewish descent but is also observed in some Hispanic populations, likely owing to historical gene flow between these two populations in Europe and America. [188]

Previously cited studies have detected evolutionary changes in coding and non-coding regions of BRCA1/2 genes within primate populations, and to extend the number of predicted amino acid changes that would affect gene function. [177] One such study found high-risk elements to be remarkably stable in hominoid primates, having been conserved in chimpanzee, gorilla, orangutan, and rhesus macaque. [178] As the Pavlicek study noted, the majority of insertion mutations took place in the ancestral lineage leading to hominoid primates after the split of Hominidae (25–14 MYA) and the rhesus macaque branch; more recent hominoid lineages acquired mostly deletions. [178] Furthermore, it was shown that disproportionately lower incidences of breast and ovarian cancer occur in closely related hominids, such as chimpanzees, our closest living relatives. [176]

Figure 17 and Figure 18 both illustrate mRNA BRCA1/2 cancer susceptibility sequences nested within the branches of intergroup lineages or clades, in accordance with each respective species. In all cases, we find that cancer susceptibility variants precede a divergent event between two different species; outgroup taxons are represented by non-cancer variants, in each vertical instance (see Figure 17 & Figure 18). These results depart from the conventional interpretation of cross-species inheritance of a single mutation type and support the conclusions reported by others noted previously. My findings further suggest that specific mutation types leading to altered gene expression in error-free DNA repair mechanisms vary widely among separate taxonomical rankings. We can then infer from these results that cancer-causing alterations appear to originate within localized gene pools at separate junctures throughout evolutionary time.

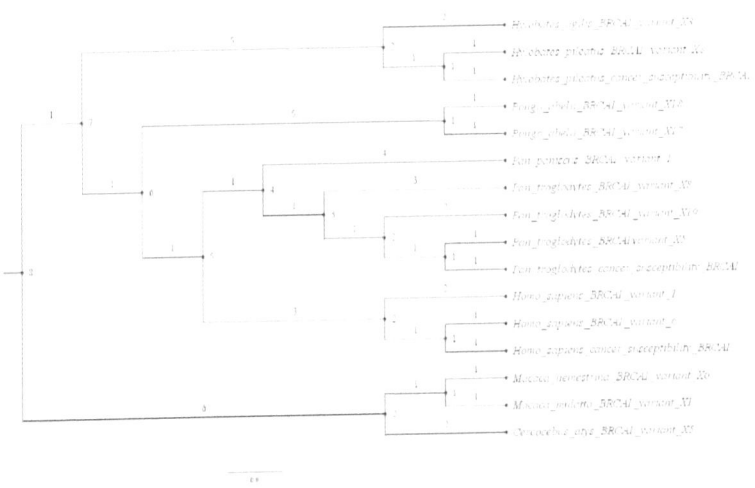

Figure 17. Phylogenetic Tree of Sixteen Primate mRNA BRCA1 cancer susceptibility and non-cancer susceptibility sequences; including: *Cercocebus atys, Homo sapiens, Hylobates agilis, Hylobates pileatus, Macaca nemestrina, Macaca mulatta, Pan paniscus, Pan troglodytes, Pongo abelii.*

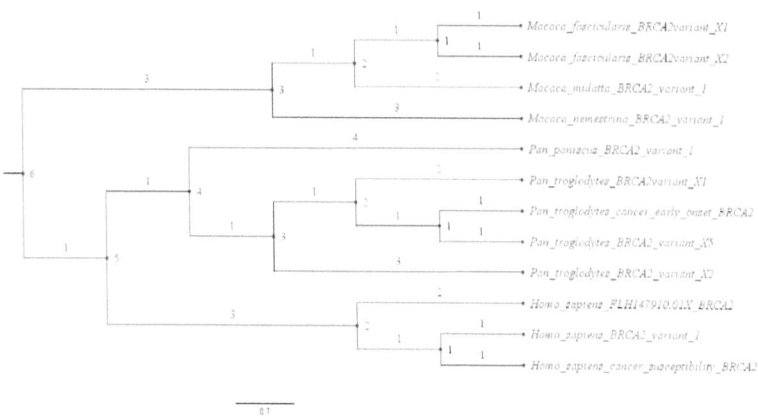

Figure 18. Phylogenetic Tree of Twelve Primate mRNA BRCA2
cancer susceptibility and non-cancer susceptibility sequences;
including [in alphabetical order]: *Homo sapiens, Macaca fascicularis,
Macaca nemestrina, Macaca mulatta, Pan paniscus, Pan troglodytes.*

Pairwise Identity Ratios

To align a pair of sequences, a scoring system is required to
score matches and mismatches. [189] This helps identify variation
in rates and degrees of dissimilarity between two or more sets
of homologous sequences. Variation itself is a consequence of
different factors, including the mutation process, genetic drift,
and natural selection. Substitutions, insertions, and deletions may
occur at different rates over evolutionary time. [190]

For mRNA BRCA 1/2 sequences, the relative rates of different
substitutions can be empirically determined by comparing each
sequence to a specific biomarker; *Homo sapiens* was chosen in this
individual instance. These empirical measurements can then form
the basis for determining various degrees of evolutionary change.
Here, pairwise alignment revealed an unusually high proportion
of dissimilarities between cancer susceptibility sequences among
members of each distinct group or species. Overall, pairwise
identity in the aligned segments ranged from 93% to 99% for

non-cancer variants. Identity ratios dropped from 76% to 96% when cancer susceptibility or early onset sequences were included (see Table 7).

Annotation Number	BRCA1	BRCA2
NR_027676.1 *Homo sapiens*	0.99	-
NM_007294.3 *Homo sapiens*	0.99	-
AF005068.1* *Homo sapiens*	0.79	-
FLH147910.01X *Homo sapiens*	-	0.99
NM_000059.*3* *Homo sapiens*	-	0.98
U43746.1* *Homo sapiens*	-	0.96
XM_009432096.2 *Pan paniscus*	0.98	-
XM_016930482.1 *Pan troglodytes*	0.91	-
XM_016930483.1 *Pan troglodytes*	0.91	-
NM_001301758.1* *Pan paniscus*	0.77	
XM_003826866.2 *Pan paniscus*	-	0.98
XM_016925134.1* *Pan troglodytes*	-	0.91
XM_019039974.*1* *Pan troglodytes*	-	0.97
XM_002827484.2 *Pongo abelii*	0.97	-
XM_009251713.1* *Pongo abelii*	0.79	-
KM017622.1* *Pongo pygmaeus*	0.76	-

XM_002824150.*2* *Pongo abelii*	–	0.97
XM_003913740.*3* *Macaca nemestri*na	–	0.95
XM_005585600.*2* *Macaca fasciclari*s	–	0.95
XM_011748633.1* *Macaca nemestri*na	–	0.89
XM_015119744.*1* *Macaca mulat*ta	0.94	–
XM_011725007.1 *Macaca nemestri*na	0.94	–
XM_003270250.*3* *Nomascus leucogeny*s	–	0.97
XM_010367067.*1* *Rhinopithecus roxell*ana	–	0.95
XM_012064631.*1* *Cercocebus aty*s	–	0.95
XM_011997808.*1** *Mandrillus leuchopha*eus	–	0.91
XM_003279499.1 *Nomascus leucogeny*s	0.97	–
XM_012046793.1 *Cercocebus aty*s	0.95	–
KM017619.1* *Hylobates pileatus*	0.76	–
KM017620.1* *Hylobates agilis*	0.76	–
KM017626.1* *Symphalangus syndactylus*	0.76	–

Table 7. Pairwise comparison of 10 primate family
mRNA BRCA1/2 sequences against *Home sapiens*

* represents cancer susceptibility or early onset sequences

Consensus Results of Breast Cancer 1 Early
Onset BRCA1 mRNA Partial cds

Pairwise comparisons reveal quantifiable degrees of variation between two sets of sequences: namely, non-cancer variants versus cancer susceptibility. To further examine the question of statistical degrees of genetic alteration within a species, consensus values were generated between two or more individual sequences in human populations. As highlighted in Table 8, proportions of dissimilarity in *Homo sapiens* (BRCA1) ranges from 21% to 24%. Incidentally, these discrepancies are generally found to be lower in other primates (ranging from 7% to 22%), excluding *Macaca mulatta* (BRCA1). Coupled with the results from pairwise comparison, consensus results show *Homo sapiens* having the highest proportion of dissimilarity among the entire dataset of primate families.

Annotation Number	Consensus %
NM_007294.3 *non-cancer variant*	0.99
AY890755.1*	0.76
AY888184.1*	0.76
U14680.1*	0.78
AB385129.1*	0.76
AY751490.1*	0.75
AF005068.1*	0.77

Table 8. Consensus analysis of BRCA1 cancer susceptibility against non-cancer variant *(Homo sapiens)*

* represents cancer susceptibility or early onset sequences

Summary

On various levels, interspecies comparisons reveal the existence of unexpectedly high degrees of variation among the selected mRNA BRCA1/2 sequences. As the results from phylogenetic reconstruction demonstrate, cancer susceptibility variants are distinct to their respective clades and do not occur on outgroups of other species. Predisposition toward tumor suppression crosses taxonomical boundaries, but cancer-causing alterations in BRCA1/2 appear to originate within localized gene pools at separate junctures throughout evolutionary time. This supports the explanation that BRCA1/2 genes may be undergoing rapid evolution, as revealed by unusually high proportion of dissimilarities between cancer susceptibility sequences among members of each distinct group or species; *Homo sapiens* having the highest proportion of dissimilarity among the 10 primate families. The genomic datasets used in the study were limited relative to larger, more comprehensive investigations. Further examination is needed to validate my findings.

BONUS CHAPTER

9

FOLLOW THE FACTS, WHEREVER THEY MAY LEAD

Originally published for Diaries of Dissension, iUniverse 2011

nce I read a story about a beautiful garden. In this garden lived a man named Adam and a woman named Eve. Adam and Eve shared this garden with all the creatures of the Earth. Created from the dust of the ground, Adam alone was chosen by God to be his mirror image. He was given full dominion over all living things—the fish of the sea, the fowl of the air, the cattle, and every creeping thing upon the Earth.

In the midst of this garden lay the tree of life and the tree of the knowledge of good and evil. God commanded Adam and Eve to eat freely from any of the trees in the Garden, with one exception. It was forbidden to eat from the tree of the knowledge of good and evil. Eve, who was made of a rib taken from Adam's body, was tempted by the serpent in the Garden to eat the fruit of the forbidden tree. Adam also ate from the tree. Now Adam and Eve possessed the knowledge of good and evil, like God himself.

God became aware of their disobedient ways and grew enraged. To show his disappointment, he cursed the serpent to crawl on its belly forever and cursed Eve by multiplying the pain of birth. God then drove Adam and Eve from the Garden of Eden, never to return. And from these beginnings, all men and women on Earth have descended.

I am sure you are somewhat familiar with this tall tale. And if you are not, here is your first lesson in Old Testament mythology. The above mentioned is representative of the first section of Creation from the book of Genesis. In particular, I was attempting to describe God's creation of man and his role over all living things on Earth, according to biblical accounts. Like any other creation story of its kind, it too shares strikingly similar features to other myths commonly found throughout human culture. Often set in a nonspecific past, most creation myths describe life's earliest beginnings and humanity's role in the world. They all seem to have elements of good and evil, right and wrong. Man is usually portrayed as the centerpiece. In much the same way, Adam was created in God's image and given dominion over all the Earth. Creation myths typically involve a plot and characters who are either gods, humanlike figures, or animals that often speak or transform easily. [191] This early account of Genesis fits right into that description.

Indeed, even the most hard-core believer must admit at some level that certain elements of Genesis sound rather fictional. I know many moderate Christians who lean in this direction but simply fail to admit it outright. This is almost certainly a byproduct of their own fear, guilt, or ignorance, perhaps. Rather than rejecting it the on the conclusions of their own logical reasoning, they accommodate these biblical fictions with a label of symbolism and metaphor. Unfortunately for them, the Bible does not come with a set of instructions telling its readers what is to be taken literal and what to be taken metaphorically. This is one fundamental flaw that is conveniently overlooked by believers of the Abrahamic religions.

I find it quite troublesome comprehending how anyone can possibly accept the accounts of Genesis as literal truth, especially grown adults. But on the far end of the spectrum reside those Christians who strongly support the idea that the entire Bible is a historically accurate record of actual events. In fact, according to several polls, it is estimated that up to 45 percent of American adults take a literal interpretation of the Genesis creation narrative. [192] Frankly, to believe this, you must be willing to accept some farfetched notions.

For one, according to biblical records, the Earth [and universe combined] is estimated to be between five thousand and ten thousand years of age. Biblical scholars and theologians base these figures on the genealogies of Genesis and other dates in the Bible. [193] On the other hand, scientifically acquired estimates tell us quite a different story about the age of the Earth. Using the latest geochronological dating methods, such as radiometric dating or chemostratigraphy, geological figures point to an approximate age of 4.54 billion years (according to the US Geological Survey). [193] When we compare biblical estimates to scientific approximates, something obvious quickly jumps at you — the disparity between the two estimates is quite wide. This is not just by a couple of thousand years, or a couple of million even, but a staggering ratio of more than 99.99999999+ percent. Well, there goes the red flag. Now ask yourself, why would the biblical God create the world just six thousand years ago, yet make it look far older?

Perhaps most crucial in this discussion is a theory in biology that explains the change, diversification, and distribution of living organisms over extended periods of geological time. In fact, throughout the course of this book, we have been discussing it all along. Supported by overwhelming evidence from genetics, paleontology, embryology, biochemistry, and more, this explanation is the cornerstone of all modern biology. In biological circles, we refer to this as the theory of evolution.

Defining biological evolution

Evolution is the change in the inherited traits of a population of organisms through successive generation. After a population splits into smaller groups, these groups evolve independently and may eventually diversify into new species. A nested hierarchy of anatomical and genetic similarities, geographical distribution of similar species and the fossil record indicate that all organisms are descended from a common ancestor through a long series of these divergent events, stretching back in a tree of life that has grown over the 3,500 million years of life on Earth. [194]

Key mechanisms in evolution

Natural selection is the gradual, nonrandom process by which biological traits become either more or less common in a population as a function of differential reproduction of their bearers. It is a key mechanism of biological evolution. Variation exists within all populations of organisms. This occurs partly because random mutations cause changes in the genome of an individual organism, and these mutations can be passed to offspring. Throughout the individuals' lives, their genomes interact with their environments to cause variations in traits. The environment of a genome includes the molecular biology in the cell, other cells, other individuals, populations, species, as well as the abiotic environment. Individuals with certain variants of the trait may survive and reproduce more than individuals with other variants. [194]

Observations of the natural world show that living organisms have undergone modification in the past and still do so in present day. The extent of the fossil record reveals morphological similarities between organisms alive today and those that once existed. Geological records coupled with the fossilized remains of extinct species of plants and animals give us insights into evolutionary time scales, relationships, distribution, and the diversity of all the living things. The genetic record of living organisms further supports the notion of relatedness and common ancestry, which are further demonstrated in the case studies found within the pages this book. Species that are closely related share a higher degree of genetic similarity than species that are further apart on evolutionary scales. What appears very certain in this discussion is that living organisms were *not* independently created [as biblical accounts would imply] but have descended and diversified over geological time from common ancestors.

Genetics, Complexity, and the Origins of Life

It should then be emphasized, biological evolution does not describe the origins of life, but rather the change, diversification, and distribution of species over geological time; and it explains the mechanisms involved in those changes. Inquiries regarding the origins of life are complicated to resolve and cross over into other disciplines of science. There is even an independent branch of studies called abiogenesis that proposes several plausible explanations for the origins of life. Some scientists speculate as to whether it could have involved the gradual chemical evolution of molecules in ocean waters near deep-sea vents.

Several experiments in science have simulated early Earth conditions and tested for spontaneous chemical reactions leading to the precursors of organic compounds. The most well-known is the Miller-Urey experiment. Results from the 1953 Miller-Urey experiment generated nine distinct amino acids from inorganic material. [195] Recent reanalysis of the samples produced during the Miller-Urey revealed that a wider range of amino acids was actually present. Amino acids are the building blocks of protein, and protein is made of the material in DNA. [196]

DNA is composed of five ordinary elements: hydrogen, oxygen, nitrogen, carbon, and phosphorus. A combination of these elements is bound together in strings of nitrogenous base pairs and held intact by phosphate ends to make a double helix. This is the basic structure for DNA. But the chemical substances found in DNA molecules are also present in inorganic substances. Furthermore, the chemistry for life is abundant throughout the cosmos, not limited to Earth or people or living things.

The question of design now comes into play. Of course, no one will argue against the complexity of genetic structures. DNA molecules are extraordinarily intricate and complex, and unlike any molecular structure we have seen thus far. Much is often said about the complexity of the cell, and I would agree. But does complexity require intelligence? For instance, a spider web is complex. Is a spider intelligent for creating a spider web? Nature produces near-perfect symmetry without externally guided influences. Structures like crystalline, snow crystals, and self-organized convection cells look so beautifully intricate, they almost appear to have the unmistakable mark of intent. Yet, for all their beauty and variation, they simply emerge as byproducts of natural processes. Thus, complex phenomena can emerge without intelligence.

I often hear creationists raise the question, can *coding systems* exist without design? And this brings me to my next point. Does complexity in living systems resemble a product of "intelligent design"? Some creation advocates would argue, yes. But this is a rule of thumb: design aims for simplicity, not complexity. For example, think of the engineers who design programming code for computer applications. The very basic principles of computer programming are based on the premises of simplicity and organization. Programming code that is simple and concise tends to be more efficient and reliable. [197] Data flow relies on the precision of efficient programs. [198] Ideally, the more simplified the code, the more efficient it may be. And the reciprocal of that statement is also true. Overly complex coding schema tends to be saturated, having unfavorable results in both processing and execution. Simplicity removes the unessential through thoughtful reduction. "Simplicity and avoiding complexity should always be a key goal." [199] Again, these are basic protocols of programming for computer applications.

Generally speaking, there are functional benefits to designing for simplicity. One can apply the same principles to many systems and designs of engineering. It should be noted, nonetheless, simplicity is not a requirement of design, only an attribute of an *efficient* design. Now, looking at living systems, one cannot ignore the obvious — biological systems are vastly inefficient and cannot account for a great amount of waste. We can examine countless examples of this in living systems. For one, I often hear people marvel at the machinery of the human anatomy, quite frequently giving it the label of being "the perfect machine." But if this was truly the case, then why bother ourselves with doctors, physicians, or health-care systems? For the sake of this discussion, I will point to one particular instance because it deals directly with the issue of DNA. Cancer occurs as a byproduct of somatic mutations in the genome of a cell. Cancer can be caused by things like radiation, chemicals, infections, genetic factors, or spontaneous errors during cell replication. One reliable source defines cancer as the following: "Cancer is essentially a disease of mitosis — the normal 'checkpoints' regulating mitosis are ignored or overridden by the cancer cell. Cancer begins when a single cell is transformed or converted from a normal cell to a cancer cell."[200] Quite simply, errors in DNA replication are the basis of malignant changes. All cancers begin in cells, the most basic unit of life. Once cancer cells get going, they can divide and grow uncontrollably, causing harm to the host organism. About 13 percent of all human deaths each year are cancer related, afflicting young children and people of all ages.[201] In light of these error-riddled missteps that occur during cell replication, ask yourself once again, do characteristics such as these resemble a product of intelligent design?

I believe the issue in question is not so much whether coding systems derive from intelligence, but whether poor results constitute intent or design. Moreover, biological systems appear more having been shaped and molded by evolutionary processes and trade-offs than by the implementation of a lackluster designer. I will get into how evolutionary mechanisms shape and mold biology as we go along.

So, it seems that life itself lacks planning, genetic information is self-constructed in environmental context, and biological systems are vastly complex. We have also seen that nature can achieve complexity on many different scales. New techniques in biotechnology are now allowing us to modify strands of DNA in order see what early genomic structures may have resembled. Scientists are also able to observe the effects of amino acids under early primordial conditions. This could help us determine how early organic molecules may have functioned and evolved.

The Role of Genetic Mutation in Evolution

Mutations occur anywhere in the body and can be beneficial, neutral, or harmful to the host organism(s). As noted earlier, cancer is an example of a harmful type of genetic mutation. Other somatic mutations are so slight that they do not cause any significant harm to the host organism. Although most mutations are naturally occurring, not all mutations are necessary to biological evolution. [202]

Before proceeding, let us consider the following. First, regardless of what philosophical position you take, I think most people would agree that DNA ultimately determines the physical and behavioral attributes of an individual organism [or a group of organisms]. DNA can be described as a genetic fingerprint, and it is arranged for biological function. So, would not major recombinant changes to DNA affect physical and behavioral appearance? Here is something else to consider: we know that mutations occur in the DNA of sexual reproductive cells. These are the types of cells that are important in sexual reproduction and that play a vital role in heredity. Passed from one generation to the next, germ-line mutations usually create tiny changes to the phenotype of an organism. Ask yourself, what were to happen when tiny inheritable mutations accumulate over tens of thousands of generations?

Coincidentally, we already have a direct observation in response to that question. A recent experiment conducted by evolutionary biologists Zachary Blount and Richard Lenski of Michigan State University put twelve genetically identical E.coli populations in complete isolation and subjected each to various types of environmental pressures over a span of twenty-five years and fifty thousand *E. coli* generations. [203] The result: each *E. coli* population developed its own evolutionary adaptations, including an increase in cell size, shape variation, growth rate, and behavior, and one population even evolved the ability to grow on citric acid. [203] A wide array of genetic change was reported, and four of the twelve populations lost the ability to repair DNA. Biological evolution was observed in real time.

Bacteria use horizontal gene transfer as one primary method for exchanging genetic information. Also, not all organisms evolve at the same rate. The rapid replication rates observed in bacteria differ substantially from the reproduction patterns of larger multicellular organisms. One might observe large-scale genetic change among bacterial populations over the course of a human lifetime, yet this becomes rather difficult to quantify when speaking in terms of multicellular eukaryotes. This is where the fossil record often helps us fill in the blanks. Nonetheless, the Blount and Lenski experiment demonstrates how environmental variables and genetic isolation can cause one genetically distinct organism to split/modify into other versions of the original. This also occurs in nature, on grandeur scales. Given enough time and geographical isolation, genetically identical populations of species can deviate, split, and develop distinct features that make them "look and behave different" over large-scale time periods, and eventually descend into more genetically distant categories. This naturally occurring process is known as speciation. And thus, natural selection acting upon genetic mutation causes these evolutionary changes to occur.

Defining speciation

Speciation is the evolutionary process by which new biological species arise. There are four geographic modes of speciation in nature, based on the extent to which speciating populations are geographically isolated from one another: allopatric, peripatric, parapatric, and sympatric. Isolated populations undergo genotypic and/or phenotypic divergence as: (a) they become subjected to dissimilar selective pressures; (b) they independently undergo genetic drift; (c) different mutations arise in the two populations. When the populations come back into contact, they have evolved such that they are reproductively isolated and are no longer capable of exchanging genes. [204]

In summary, one individual species does not simply "evolve" into another species (a common misunderstanding of biological evolution), but rather splits into two or more genetically distinct populations.

Evidence from the Fossil Record

The most well understood evidence in favor of evolution comes from the preserved remains of plants and animals. Fossils found throughout the planet are arranged in sequential order within horizontal sedimentary rock deposits. Sedimentary rock deposits are formed by layers of silt and mud that accumulate over geological time. Each layer of rock represents a specific epoch in Earth's past. The lower layer contains older rock. The upper layers contain newer rock. The earliest fossils can be found at the lower layers, while the more recent fossils are found in the upper layers. [205]

When we examine the fossil record, we see a logically consistent distribution of species over geological time. And as we further investigate each layer of rock, or strata, we see change and extinction occurring at every layer, at each epoch. We find earlier forms of species that exhibit primitive traits in comparison to new ones found in later time periods. Even in precise sequence do we see the physical transition of various insects, spiders, fish, whales, birds, horses, and humans. Despite the rarity of fossilization, approximately over 250,000 fossil species have been discovered and documented. [206]

The fossil records point to clear morphological similarities between groups of plants and animals. For example, comparisons of pelvic bone structure show how modern birds are descendants of dinosaurs. The structural patterns of limb bones indicate transitions between amphibians and mammals. These limb bones can be traced back to the first amphibians that walked. All mammals today share common bone structures to their biological ancestors. Like them, we too have humerus bones, wrist bones, palm bones, and digits. Through evolutionary processes, these fundamental structures have been modified to fit different environments. In monkeys, the forelimbs became elongated to form grasping hands used for climbing and swinging from trees. A mole has a pair of short limbs for burrowing. An anteater has an enlarged third digit for tearing down ant hills and termite nests. [207] And so on, and so on . . .

Although the fossil record does not include every plant and animal that ever lived, it does provide direct evidence of common ancestry between and among different populations of organisms. It also gives us a glimpse into the geological timeframes required for understanding the evolution of life on Earth.

Confirmation from Modern Genetics

In the end, no evidence for biological evolution is more compelling than that coming from modern genetics. In many biotech laboratories, comparisons of genomic sequences are used for characterizing functional regions in DNA and identifying operations in cellular processes that help diagnose, treat, and even prevent several diseases that affect human beings. [208] As it relates to the topic at hand, comparative genomics also serves other practical purposes. For some years now, scientists have been using comparative genomics to outline the similarities of evolutionary relationships between species and groups of species. Today, computer-based bioinformatics applications such as BLAST are used to search collaborative archives of raw genetic data. Genome databases, like those mentioned in previous chapters, now house over 190 billion nucleotides of 260,000 organisms, including that of humans. [209]

As described in chapter 2, distance-matrix methods coupled with Bayesian inference algorithms are incorporated with computational biology software that reconstruct homologous colinearity among related genomes to produce phylogeny data matrices of multiple sequence alignments. The resulting phylogenies depict diagrams that illustrate branching structures that look like "trees" containing roots and nodes. Each node on the tree represents a taxonomic unit, and each interior node represents a common ancestral unit in which latter branches diverge. By pairing sequence similarities versus that of genetically distant organisms, it is possible to produce high-degree diagrams outlining evolutionary relatedness amongst different populations of species. And though internal nodes cannot be directly observed in nature, they are, however, further confirmed by morphological similarities found in the fossil record in direct comparison to that of modern species.

Embedded within the clade of modern humans and other great apes are unmistakable marks of African origin. Previous chapters have outlined the genetic relationships between chimpanzees, gorillas, orangutans, and gibbons, and it suggests divergence with those lineages and other common ancestors sometime in the past. Human DNA sequences are approximately 1.8 percent divergent from those of our nearest genetic relative, the chimpanzee, and 2.3 percent from gorillas. [210] Additionally, humans and chimps share seven different instances of virogenes, while all other primates share retroviruses congruent with resulting phylogenies. [210]

Direct evidence for common lineages between humans and other members of the great ape family is also found in human chromosome 2. For instance, humans have twenty-three pairs of chromosomes, while chimps, gorillas, and orangutans have twenty-four. [211] Chimpanzees have near-identical DNA sequences to human chromosome 2, but they are found in two separate chromosomes. The same is true of the more distant gorilla and orangutan. [212] It is often explained that human chromosome 2 is a result of an end-to-end fusion of two ancestral chromosomes that resulted in the correspondence of chromosome 2. [212] In other words, humans have one fewer chromosome pair in their cells than apes, due to a mutation found in chromosome number 2 that caused two chromosomes to fuse into one. [212] The fusion between two ancestral chromosomes therefore presents very strong evidence in favor of a common ancestor between humans and other members of the great ape family.

Taking a much closer look, human DNA reveals even more startling insights into our relationships to one another. In 2003, scientists at the Human Genome Project announced that they had mapped the entire human genome. This was a significant point in modern science. For the first time, scientists were able to reconstruct phylogenies based on the genomic data belonging to different human populations. Of the three billion base pairs in the human genome, one type has significance in determining relatedness among immediate populations — mitochondrial DNA (mtDNA). [213]

Mitochondrial DNA is inherited solely from the mother, as noted numerous times. This form of DNA is found to mutate more rapidly than other molecules like proteins or rRNAs, which tell us more about ancient evolutionary relationships. Mitochondrial DNA is not affected by the process of natural selection, allowing mutations to accumulate more freely in noncoding DNA. Using the latest techniques in genome sequencing, scientists have acquired hundreds of thousands of genomic samples from human beings all over the globe. Several thousand human volunteers, representing distinct populations from all parts of the world including a dozen different tribal groups in Africa were subjected to genetic testing.[214]

In almost all cases, two or more people sharing the same sequence also belonged to the same population. Of the one hundred different sequences identified, the greatest diversity of sequences was found among the various African populations. [214] The data would later reveal a "human family tree" by which all present-day human sequences could be narrowed down to a single ancestral sequence. Computational phylogenetics of human populations indicate an estimated time and point of origin — East Africa, about 80,000-200,000 years ago. [213 214 215 216 217]

Darwin was Right About Common Descent

I opened this book referencing one of Darwin's great achievements. It is only proper that we end with his most profound implications. One of the single most important concepts we have learned is that every species belongs to a parent clade on a branch, regardless of where it resides on the tree of life. There is not a single lineage that arouse independently outside of some nested root within. Simply based on this remarkable insight, there is an overwhelming consensus among people of science. In fact, it is hardly a debate. According to a recent Gallup poll, over 95 percent of all world scientists accept evolution as the most valid set of principles that best explains the diversity of life. One news article published in *Newsweek* said the following: "By one count there are some 700 scientists with respectable academic credentials out of a total of 480,000 U.S. earth and life scientists who give credence to the counter evolutionary argument, also known as creation science." [218] Something worth noting: belief in creationism is inversely correlated to education. Of those with postgraduate degrees in the United States, 74 percent accept evolution as a valid explanation, rather than the alternative claim brought forth by critics of evolutionary theory. [218]

I think it is worth pointing this out. Misconceptions are prevalent. Most of it being the product of lies fueled by Christian elements and other creation advocate groups. Furthermore, I cannot presume to call myself an educator, and just stand by and do nothing about it. Let me make this clear; all independent disciplines of science converge on this issue. Dr. Richard Dawkins said it best: "Today the theory of evolution is about as much open to doubt as the theory that the earth goes round the sun." [219]

So, where does that leave us? Charles Darwin was right about common ancestry, despite what you might have heard. Naturally, the repercussions stemming from evolutionary studies stretch far beyond scientific circles and into congregations, churches, and mosques. Biological evolution all but exposes the illegitimate claim of divine creation. Darwin himself understood this too well. He knew that while his theory of evolution, powered by the mechanism of natural selection, produced a good explanation of adaptation, it also removes the need for "intelligent design".

And thus, species of living organisms were not independently created, but have descended and diversified over time from common ancestors. To date, no other scientific theory better explains this phenomenon. Evolutionary theory has withstood the test of time — by way of vicarious experimentation, observation, analysis, and relentless criticism; though, opposing viewpoints still cling to the concept of "intelligent design". As a researcher and life science educator, I cannot subscribe myself to such misguided notions that suggest static biological states. Surely, proper examination of the natural world reveals evolutionary trajectories — some random, others nonrandom — and all having observable genetic implications. It is only when we apply evolutionary explanations to living systems that it becomes ever so clear. The world was not specifically designed with us in mind, but rather we long since adapted and conformed to our surroundings, only giving it the illusionary appearance of "design".

NOTES

1   Darwin, C., Chancellor, G. R., & Van Wyhe, J. (2009). Charles Darwin's notebooks from the voyage of the Beagle. Cambridge, England: Cambridge University Press.

2   Darwin, C. (2004). On the origin of species, 1859. Routledge.

3   Gregory, T. R. (2008). Understanding evolutionary trees. Evolution: Education and Outreach, 1(2), 121-137. University of California, Berkeley.

4   Daugelaite, J., O'Driscoll, A., & Sleator, R. D. (2013). An overview of multiple sequence alignments and cloud computing in bioinformatics. International Scholarly Research Notices, 2013.

5   Lassmann, T., & Sonnhammer, E. L. (2005). Kalign-an accurate and fast multiple sequence alignment algorithm. BMC bioinformatics, 6(1), 298.

6   Felsenstein, J. (1981). Evolutionary trees from DNA sequences: A maximum likelihood approach. Journal of molecular evolution, 17(6), 368-376.

7   Hurst, G. D., & Jiggins, F. M. (2005). Problems with mitochondrial DNA as a marker in population, phylogeographic and phylogenetic studies: the effects of inherited symbionts. Proceedings of the Royal Society B: Biological Sciences, 272(1572), 1525-1534.

8   Sequeira, F., Sodré, D., Ferrand, N., Bernardi, J. A., Sampaio, I., Schneider, H., & Vallinoto, M. (2011). Hybridization and massive mtDNA unidirectional introgression between the closely related Neotropical toads Rhinella marina and R. schneideri inferred from mtDNA and nuclear markers. BMC evolutionary biology, 11(1), 1-15.

9   Miller, W., Schuster, S. C., Welch, A. J., Ratan, A., Bedoya-Reina, O. C., Zhao, F., & Tomsho, L. P. (2012). Polar and brown bear genomes reveal ancient admixture and demographic footprints of past climate

change. Proceedings of the National Academy of Sciences, 109(36), E2382-E2390.

10 Nolte, A. W., & Tautz, D. (2010). Understanding the onset of hybrid speciation. Trends in Genetics, 26(2), 54-58. DOI: 10.1016/j.tig.2009.12.001

11 Nishihara, H., Satta, Y., Nikaido, M., Thewissen, J. G. M., Stanhope, M. J., & Okada, N. (2005). A retroposon analysis of Afrotherian phylogeny. Molecular biology and evolution, 22(9), 1823-1833.

12 Orlando, L., Hänni, C. and Douady, C.J. (2007) Mammoth and Elephant Phylogenetic Relationships: Mammut Americanum, the Missing Outgroup. Evolutionary Bioinformatics Online, 3, 45.

13 de Jong, W.W., Zweers, A. and Goodman, M. (1981) Relationship of Aardvark to Elephants, Hyraxes and Sea Cows from α-Crystallin Sequences. Nature, 292, 538-540.

14 Honeycutt, R.L. (2008) Small Changes, Big Results: Evolution of Morphological Discontinuity in Mammals. Journal of biology, 7, 9.

15 Rohland, N., Reich, D., Mallick, S., Meyer, M., Green, R.E., Georgiadis and Hofreiter, M. (2010) Genomic DNA Sequences from Mastodon and Woolly Mammoth Reveal Deep Speciation of Forest and Savanna Elephants. PLoS Biology, 8, Article ID: e1000564.

16 Chan, Y. C., Roos, C., Inoue-Murayama, M., Inoue, E., Shih, C. C., Pei, K. J. C., & Vigilant, L. (2010). Mitochondrial genome sequences effectively reveal the phylogeny of Hylobates gibbons. PLoS One, 5(12), e14419.

17 Hurst, G.D. and Jiggins, F.M. (2005) Problems with Mitochondrial DNA as a Marker in Population, Phylogeographic and Phylogenetic Studies: The Effects of Inherited Symbionts. Proceedings of the Royal Society B: Biological Sciences, 272, 1525-1534.

18 Salmonella (non-typhoidal) (2013) World Health Organization.

19 Beceiro, A., Tomás, M., & Bou, G. (2013). Antimicrobial resistance and virulence: a successful or deleterious association in the bacterial world?. Clinical microbiology reviews, 26(2), 185-230.

20 Forsberg KJ, Reyes A, Wang B, Selleck EM, Sommer MO, et al. (2012) The shared antibiotic resistome of soil bacteria and human pathogens. Science 337: 1107-1111.

21 Liu, B., & Pop, M. (2009). ARDB—antibiotic resistance genes database. Nucleic acids research, 37(suppl_1), D443-D447.

22 Kapoor, G., Saigal, S., & Elongavan, A. (2017). Action and resistance mechanisms of antibiotics: A guide for clinicians. Journal of anaesthesiology, clinical pharmacology, 33(3), 300.

23 Peterson, E., & Kaur, P. (2018). Antibiotic resistance mechanisms in bacteria: relationships between resistance determinants of antibiotic producers, environmental bacteria, and clinical pathogens. Frontiers in microbiology, 9, 2928.

24 Nesme, J., & Simonet, P. (2015). The soil resistome: a critical review on antibiotic resistance origins, ecology and dissemination potential in telluric bacteria. Environmental microbiology, 17(4), 913-930.

25 Manyi-Loh, C., Mamphweli, S., Meyer, E., & Okoh, A. (2018). Antibiotic use in agriculture and its consequential resistance in environmental sources: potential public health implications. Molecules, 23(4), 795.

26 Blair, J. M., Webber, M. A., Baylay, A. J., Ogbolu, D. O., & Piddock, L. J. (2015). Molecular mechanisms of antibiotic resistance. Nature reviews microbiology, 13(1), 42-51.

27 Cox, G., Stogios, P. J., Savchenko, A., & Wright, G. D. (2015). Structural and molecular basis for resistance to aminoglycoside antibiotics by the adenylyltransferase ANT (2 ″)-Ia. MBio, 6(1).

28 Perron, G. G., Whyte, L., Turnbaugh, P. J., Goordial, J., Hanage, W. P., Dantas, G., & Desai, M. M. (2015). Functional characterization of bacteria isolated from ancient arctic soil exposes diverse resistance mechanisms to modern antibiotics. PLoS One, 10(3), e0069533.

29 Schloss, P. D., & Handelsman, J. (2005). Introducing DOTUR, a computer program for defining operational taxonomic units and estimating species richness. Applied and environmental microbiology, 71(3), 1501-1506.

30 Parry, C. M. (2003). Antimicrobial drug resistance in Salmonella enterica. Current opinion in infectious diseases, 16(5), 467-472.

31 Chaguza, C., Cornick, J. E., & Everett, D. B. (2015). Mechanisms and impact of genetic recombination in the evolution of Streptococcus pneumoniae. Computational and structural biotechnology journal, 13, 241-247.

32 Davies, J., & Davies, D. (2010). Origins and evolution of antibiotic resistance. Microbiology and molecular biology reviews, 74(3), 417-433.

33 Kraemer, S. A., Ramachandran, A., & Perron, G. G. (2019). Antibiotic pollution in the environment: from microbial ecology to public policy. Microorganisms, 7(6), 180.

34 Hori, M., Shibuya, K., Sato, M., & Saito, Y. (2014). Lethal effects of short-wavelength visible light on insects. Scientific reports, 4(1), 1-6.

35 Adewoye, A. B., Lindsay, S. J., Dubrova, Y. E., & Hurles, M. E. (2015). The genome-wide effects of ionizing radiation on mutation induction in the mammalian germline. Nature communications, 6(1), 1-8.

36 Barber, R., Plumb, M. A., Boulton, E., Roux, I., & Dubrova, Y. E. (2002). Elevated mutation rates in the germ line of first-and second-generation offspring of irradiated male mice. Proceedings of the National Academy of Sciences, 99(10), 6877-6882.

37 Tatebe, K., S. J. Chmura, and P. P. Connell. "Severe radiation toxicity associated with a germline PTEN mutation." International Journal of Radiation Oncology• Biology• Physics 99.2 (2017): E620.

38 Verhofstad, N., Linschooten, J. O., van Benthem, J., Dubrova, Y. E., van Steeg, H., van Schooten, F. J., & Godschalk, R. W. (2008). New methods for assessing male germ line mutations in humans and genetic risks in their offspring. Mutagenesis, 23(4), 241-247.

39 Pray, L. (2008). DNA replication and causes of mutation. Nature Education, 1(1), 214.

40 Upton, A. C., Shore, R. E., & Harley, N. H. (1992). The health effects of low-level ionizing radiation. Annual review of public health, 13(1), 127-150.

41 Cnossen, I., Sanz-Forcada, J., Favata, F., Witasse, O., Zegers, T., & Arnold, N. F. (2007). Habitat of early life: Solar X-ray and UV radiation at Earth's surface 4–3.5 billion years ago. Journal of Geophysical Research: Planets, 112(E2).

42 Madronich, S., McKenzie, R. L., Björn, L. O., & Caldwell, M. M. (1998). Changes in biologically active ultraviolet radiation reaching the Earth's surface. Journal of Photochemistry and Photobiology B: Biology, 46(1-3), 5-19.

43 Zepp, R. G., Erickson Iii, D. J., Paul, N. D., & Sulzberger, B. (2007). Interactive effects of solar UV radiation and climate change on biogeochemical cycling. Photochemical & Photobiological Sciences, 6(3), 286-300.

44 Hessen, D. O. (2008). Solar radiation and the evolution of life. Solar radiation and human health, 123-136.

45 Samson, P. J., & Ragland, K. W. (1977). Ozone and visibility reduction in the Midwest: Evidence for large-scale transport. Journal of applied Meteorology, 16(10), 1101-1106.

46 Douglas, R. H., & Jeffery, G. (2014). The spectral transmission of ocular media suggests ultraviolet sensitivity is widespread among mammals. Proceedings of the Royal Society of London B: Biological Sciences, 281(1780), 20132995.

47 Husemann, M., Schmitt, T., Stathi, I., & Habel, J. C. (2012). Evolution and radiation in the scorpion Buthus elmoutaouakili Lourenco and Qi 2006 (Scorpiones: Buthidae) at the foothills of the Atlas Mountains (North Africa). Journal of Heredity, 103(2), 221-229.

48 D'Suze, G. (1990). Escorpiones, características, distri-bución geográfica y comentarios generales. Emergencias por animales ponzoñosos en las Américas, 65.

49 Armas, L. F. (2001). Scorpions of the Greater Antilles, with the description of a new troglobitic species (Scorpiones: Diplocentridae), in: Scorpions 2001. In M. G. A. Polis, V. Fet, & P. A. Selden (Eds.), British Arachnological Society (pp. 245-253). Burnham Beeches, Bucks, UK.

50 Srinivas, C., Kumar, A., Rai, R., Kini, J., & Kumarchandra, R. (2015). Standardization of mean lethal dose (LD50/30) of X-rays using linear accelerator (LINIAC) in albino wistar rat model based on survival analysis studies and hematological parameters. Research Journal of Pharmaceutical, Biological and Chemical Sciences, 6(5), 1215-1219.

51 Churakov, G., Sadasivuni, M. K., Rosenbloom, K. R., Huchon, D., & Schmitz, J. (2010). Rodent evolution: Back to the root. Molecular Biology and Evolution, 27(6), 1315-1326.

52 Weinreich, D. M. (2001). The rates of molecular evolution in rodent and primate mitochondrial DNA. Journal of Molecular Evolution, 52(1), 40-50.

53 Schlager, G., & Dickie, M. M. (1971). Natural mutation rates in the house mouse estimates for five specific loc and dominant mutations. Mutation Research/Fundamental and Molecular Mechanisms of Mutagenesis, 11(1), 89-96.

54 Uchimura, A., Higuchi, M., Minakuchi, Y., Ohno, M., & Yagi, T. (2015). Germline mutation rates and the long-term phenotypic effects of mutation accumulation in wild-type laboratory mice and mutator mice. Genome research.

55 Drake, J. W., Charlesworth, B., Charlesworth, D., & Crow, J. F. (1998). Rates of spontaneous mutation. Genetics, 148(4), 1667-1686.

56 Dunn, C. W., Zapata, F., Munro, C., Siebert, S., & Hejnol, A. (2018). Pairwise comparisons across species are problematic when analyzing functional genomic data. Proceedings of the National Academy of Sciences, 201707515.

57 Ninio, M., Privman, E., Pupko, T., & Friedman, N. (2007). Phylogeny reconstruction: Increasing the accuracy of pairwise distance estimation using Bayesian inference of evolutionary rates. Bioinformatics, 23(2), e136-e141.

58 Yang, Z. (2006). Computational Molecular Evolution. Oxford University Press.

59 Hillis, D. M., & Bull, J. J. (1993). An empirical test of bootstrapping as a method for assessing confidence in phylogenetic analysis. Systematic Biology, 42(2), 182-192.

60 Jenkins, D. G., & Ricklefs, R. E. (2011). Biogeography and ecology: Two views of one world

61 Monadjem, A., Taylor, P. J., Denys, C., & Cotterill, F. P. (2015). Rodents of sub-Saharan Africa: A biogeographic and taxonomic synthesis. Walter de Gruyter GmbH & Co KG.

62 Lourenco, W. R. (2001). The scorpion families and their geographical distribution. Journal of Venomous Animals and Toxins, 7(1), 03-23.

63 Blanga-Kanfi, S., Miranda, H., Penn, O., Pupko, T., & Huchon, D. (2009). Rodent phylogeny revised: Analysis of six nuclear genes from all major rodent clades. BMC Evolutionary Biology, 9(1), 71.

64 Sharma, P. P., Baker, C. M., Cosgrove, J. G., Johnson, J. E., ..., & Giribet, G. (2018). A revised dated phylogeny of scorpions: Phylogenomic support for ancient divergence of the temperate Gondwanan family Bothriuridae. Molecular phylogenetics and evolution, 122, 37-45.

65 Prendini, L., & Wheeler, W. C. (2005). Scorpion higher phylogeny and classification, taxonomic anarchy, and standards for peer review in online publishing. Cladistics, 21(5), 446-494.

66 Takezaki, N., Rzhetsky, A., & Nei, M. (1995). Phylogenetic test of the molecular clock and linearized trees. Molecular biology and evolution, 12(5), 823-833.

67 Lartillot, N., Phillips, M. J., & Ronquist, F. (2016). A mixed relaxed clock model. Phil. Trans. R. Soc. B, 371(1699), 20150132.

68 Lanfear, R., Kokko, H., & Eyre-Walker, A. (2014). Population size and the rate of evolution. Trends in ecology & evolution, 29(1), 33-41.

69 Kumar, S., & Subramanian, S. (2002). Mutation rates in mammalian genomes. Proceedings of the National Academy of Sciences, 99(2), 803-808.

70 Hodgkinson, A., & Eyre-Walker, A. (2011). Variation in the mutation rate across mammalian genomes. Nature Reviews Genetics, 12(11), 756.

71 Ellegren, H., Smith, N. G., & Webster, M. T. (2003). Mutation rate variation in the mammalian genome. Current opinion in genetics & development, 13(6), 562-568.

72 Allio, R., Donega, S., Galtier, N., & Nabholz, B. (2017). Large variation in the ratio of mitochondrial to nuclear mutation rate across animals: Implications for genetic diversity and the use of mitochondrial DNA as a molecular marker. Molecular Biology and Evolution, 34(11), 2762-2772.

73 Gantenbein, B., & Eightley, P. D. (2004). Rates of molecular evolution in nuclear genes of east Mediterranean scorpions. Evolution, 58(11), 2486-2497.

74 Fabre, P. H., Hautier, L., Dimitrov, D., & Douzery, E. J. (2012). A glimpse on the pattern of rodent diversification: A phylogenetic approach. BMC evolutionary biology, 12(1), 88.

75 Carlin, J. L. (2011) Mutations Are the Raw Materials of Evolution. Nature Education Knowledge, 3(10), 10.

76 Nöthel, H. (1987). Adaptation of Drosophila melanogaster populations to high mutation pressure: Evolutionary adjustment of mutation rates. Proceedings of the National Academy of Sciences, 84(4), 1045-1049.

77 Cordeiro, A. R., Marques, E. K., & Veiga-Neto, A. J. (1973). Radioresistance of a natural population of Drosophila willistoni living in a radioactive environment. Mutation Research/Fundamental and Molecular Mechanisms of Mutagenesis, 19(3), 325-329.

78 Ling, C. C., & Endlich, B. (1989). Radioresistance induced by oncogenic transformation. Radiation research, 120(2), 267-279.

79 Joiner, M. C. (1994). Induced radioresistance: An overview and historical perspective. International Journal of Radiation Biology, 65(1), 79-84.

80 Deryabina, T. G., Kuchmel, S. V., Nagorskaya, L. L., Hinton, T. G., & Smith, J. T. (2015). Long-term census data reveal abundant wildlife populations at Chernobyl. Current Biology, 25(19), R824-R826.

81 Husemann, M., Schmitt, T., Stathi, I., & Habel, J. C. (2012). Evolution and radiation in the scorpion Buthus elmoutaouakili Lourenco and Qi

2006 (Scorpiones: Buthidae) at the foothills of the Atlas Mountains (North Africa). Journal of Heredity, 103(2), 221-229.

82 Hsia, C. C., Schmitz, A., Lambertz, M., Perry, S. F., & Maina, J. N. (2013). Evolution of air breathing: Oxygen homeostasis and the transitions from water to land and sky. Comprehensive Physiology, 3(2), 849-915.

83 Levin-Zaidman, S., Englander, J., Shimoni, E., Sharma, A. K., & Minsky, A. (2003). Ringlike structure of the Deinococcus radiodurans genome: A key to radioresistance? Science, 299(5604), 254-256.

84 Pryke, L. M. (2016). Scorpion. Reaktion Books.

85 Smithsonian Institution. (2018). Fluorescence in Scorpions. The Walter Reed Biosystematics Unit.

86 Ray, C. (2017). The Mystery of a Scorpion's Glow. The New York Times Book of Science Questions & Answers. Penguin Random House.

87 Kloock, C. T., Kubli, A., & Reynolds, R. (2010). Ultraviolet light detection: A function of scorpion fluorescence. Journal of Arachnology, 38(3), 441-445.

88 Crow, J. F. (2000). The origins, patterns and implications of human spontaneous mutation. Nature Reviews Genetics, 1(1), 40.

89 Rahbari, R., Wuster, A., Lindsay, S. J., Hardwick, R. J., & Stratton, M. R. (2016). Timing, rates and spectra of human germline mutation. Nature genetics, 48(2), 126.

90 Cai, L., & Wang, P. (1995). Induction of a cytogenetic adaptive response in germ cells of irradiated mice with very low-dose rate of chronic γ-irradiation and its biological influence on radiation-induced DNA or chromosomal damage and cell killing in their male offspring. Mutagenesis, 10(2), 95-100.

91 Jin, S., Jiang, H., & Cai, L. (2020). New understanding of the low-dose radiation-induced hormesis. Radiation Medicine and Protection.

92 Farwick, A., Jordan, U., Fuellen, G., Huchon, D., Catzeflis, F., Brosius, J., & Schmitz, J. (2006). Automated scanning for phylogenetically informative transposed elements in rodents. Systematic Biology, 55(6), 936-948.

93 Norris, C. M., Beeck, B., Unruh, Y. C., Solanki, S. K., Krivova, N. A., & Yeo, K. L. (2017). Spectral variability of photospheric radiation due to faculae-I. The Sun and Sun-like stars. Astronomy & Astrophysics, 605, A45.

94 Vié, J.C., Hilton-Taylor, C., Stuart, S.N. Wildlife in a changing world: an analysis of the 2008 IUCN Red List of threatened species. (2009) IUCN.

95 Vaughan, T.A., Ryan, J.M., Czaplewski, N. J. Mammalogy. (2013) Jones & Bartlett Publishers.

96 Robertson, D.S., McKenna, M.C., Toon, O.B. Survival in the first hours of the Cenozoic. (2004) GSA Bulletin 116(5-6)

97 Luo, Z.X., Yuan, C.X., Meng, Q.J., et al. A Jurassic eutherian mammal and divergence of marsupials and placentals. (2011) Nature 476 (7361): 442-445.

98 Meredith, R.W., Janečka, J.E., Gatesy, J., et al. Impacts of the Cretaceous Terrestrial Revolution and KPg extinction on mammal diversification. (2011) Science 334(6055): 521-524.

99 Eizirik, E., Murphy, W.J., O'brien, S.J. Molecular dating and biogeography of the early placental mammal radiation. (2001) J Hered 92(2), 212-219.

100 Brandt, A.L., Grigorev, K., Afanador-Hernández, Y. et al Mitogenomic sequences support a north–south subspecies subdivision within Solenodon paradoxus. (2016) Mitochondrial DNA Part A, 1-9.

101 Tobe, S.S., Kitchener, A.C., Linacre, A.M. Reconstructing mammalian phylogenies: a detailed comparison of the cytochrome b and cytochrome oxidase subunit I mitochondrial genes. (2010) PloS one 5(11): e14156.

102 Corneli, P., Ryk, H.W. "Mitochondrial genes and mammalian phylogenies: increasing the reliability of branch length estimation." (2000) Molecular biology and evolution 17(2): 224-234.

103 Figuet, E., Romiguier, J., Dutheil, J.Y., et al. Mitochondrial DNA as a tool for reconstructing past life-history traits in mammals. (2014) J Evol boil: 27(5): 899-910.

104 Bohle, H.M., Gabaldón, T. Selection of marker genes using whole-genome DNA polymorphism analysis. (2012) Evol bioinform online 8: 161-169.

105 Brown, T. A. From: Chapter 16, Molecular Phylogenetics. (2002) Garland science.

106 Arnold, M.L. Natural hybridization and evolution. (1997) Oxford University Press.

107 Bi, S., Wang, Y., Guan, J., et al. Three new Jurassic euharamiyidan species reinforce early divergence of mammals. (2014) Nature 514(7524): 579-584.

108 Ragan, M.A. Phylogenetic inference based on matrix representation of trees (1992) Mol Phylogenet Evol 1(1): 53-58.

109 Ramnauth, A. Molecular Epidemiology and Evolution of Viral Pathogens. (2013) University of Edinburgh.

110 Douady, C.J., Delsuc, F., Boucher, Y., et al. Comparison of Bayesian and maximum likelihood bootstrap measures of phylogenetic reliability. (2003) Mol Biol Evol 20(2): 248-254.

111 Huelsenbeck, J.P., Rannala, B. Frequentist properties of Bayesian posterior probabilities of phylogenetic trees under simple and complex substitution models. (2004) Sys Boil 53(6): 904-913.

112 Hall, B.G., Salipante, S.J. Measures of clade confidence do not correlate with accuracy of phylogenetic trees. (2007) PLoS Comput Biol 3(3): e51.

113 O'Leary, M.A., Bloch, J.I., Flynn, J.J. et al. The placental mammal ancestor and the post–K-Pg radiation of placentals. (2013) Science 339(6120): 662-667.

114 Excoffier, L., Smouse, P.E., Quattro, J.M. Analysis of molecular variance inferred from metric distances among DNA haplotypes: application to human mitochondrial DNA restriction data. (1992) Genetics 131(2): 479-491.

115 DeSalle, R., Giddings, L.V. Discordance of nuclear and mitochondrial DNA phylogenies in Hawaiian Drosophila. (1986) Proc Natl Acad Sci 83(18): 6902-6906.

116 Hurst, G.D. and Jiggins, F.M. Problems with Mitochondrial DNA as a Marker in Population, Phylogeographic and Phylogenetic Studies: The Effects of Inherited Symbionts. (2005) Pro Biol Sci 272(1572): 1525-1534.

117 Larsen, P.A., Marchán-Rivadeneira, M.R., Baker, R.J. Natural Hybridization Generates Mammalian Lineage with Species Characteristics. (2010) Proc Natl Acad Sci 107(25): 11447-11452.

118 Krause, J., Unger, T., Noçon, A., et al. Mitochondrial Genomes Reveal an Explosive Radiation of Extinct and Extant Bears near the Miocene-Pliocene Boundary. (2008) BMC Evol Biol 8: 220.

119 Rutty, G. East Midlands Forensic Pathology Unit; Mitochondrial DNA. (2016) University of Leicester.

120 Wible, J. R., Rougier, G. W., Novacek, M. J. & Asher, R. J. (2009). The eutherian mammal Maelestes gobiensis from the Late Cretaceous of Mongolia and the phylogeny of Cretaceous Eutheria. Bull. Am. Mus. Nat. Hist. 327, 1–123.

121 Wilson, G. P. & Riedl, J. A. (2010). New specimen reveals deltatheroidan affinities of the North American Late Cretaceous mammal Nanocuris . J. Vertebr. Paleontol. 30, 872–884.

122 Benton, M. J., Donoghue, P. C. J. & Asher, R. J. in The Timetree of Life (eds Hedges, S. B. & Kumar, S.) 35–86 (Oxford Univ. Press, 2009).

123 Rubinoff, D., & Holland, B. S. (2005). Between two extremes: mitochondrial DNA is neither the panacea nor the nemesis of phylogenetic and taxonomic inference. Systematic biology, 54(6), 952-961.

124 Kliot, A., Kontsedalov, S., Lebedev, G., Brumin, M., Cathrin, P. B., Marubayashi, J. M., & Ghanim, M. (2014). Fluorescence in situ hybridizations (FISH) for the localization of viruses and endosymbiotic bacteria in plant and insect tissues. JoVE (Journal of Visualized Experiments), (84), e51030.

125 Zheng, Y., Peng, R., Kuro-o, M., & Zeng, X. (2011). Exploring patterns and extent of bias in estimating divergence time from mitochondrial DNA sequence data in a particular lineage: a case study of salamanders (Order Caudata). Molecular Biology and Evolution, 28(9), 2521-2535.

126 Ladoukakis, E. D., & Zouros, E. (2017). Evolution and inheritance of animal mitochondrial DNA: rules and exceptions. Journal of Biological Research-Thessaloniki, 24(1), 1-7.

127 Fitzgerald, D. B., Tobler, M., & Winemiller, K. O. (2016). From richer to poorer: successful invasion by freshwater fishes depends on species richness of donor and recipient basins. Global change biology, 22(7), 2440-2450.

128 Hill, J. E. 2002. Exotic fishes in Florida. Lakeline Spring, 2002, 39-43.

129 Jeschke, J. M. 2014. General hypotheses in invasion ecology. Diversity and Distributions, 20(11), 1229-1234

130 Park, D. S., & Potter, D. 2013. A test of Darwin's naturalization hypothesis in the thistle tribe shows that close relatives make bad neighbors. Proceedings of the National Academy of Sciences, 110(44), 17915-17920.

131 Strauss, S. Y., Webb, C. O., & Salamin, N. 2006. Exotic taxa less related to native species are more invasive. Proceedings of the National Academy of Sciences, 103(15), 5841-5845.

132 Tsutsui, N. D., Suarez, A. V., Holway, D. A., & Case, T. J. 2000. Reduced genetic variation and the success of an invasive species. Proceedings of the National Academy of Sciences, 97(11), 5948-5953.

133 Gupta, A., Bhardwaj, A., Sharma, P., & Pal, Y. 2015. Mitochondrial DNA-a tool for phylogenetic and biodiversity search in equines. Journal of Biodiversity & Endangered Species, 2015.

134 Moore, W. S. 1995. Inferring phylogenies from mtDNA variation: mitochondrial-gene trees versus nuclear-gene trees. Evolution, 49(4), 718-726.

135 Shafland, P. L., Gestring, K. B., & Stanford, M. S. 2008. Florida's exotic freshwater fishes—2007. Florida Scientist, 220-245.

136 Friedman, M., Keck, B. P., Dornburg, A., Eytan, R. I., Martin, C. H., Hulsey, C. D., & Near, T. J. 2013. Molecular and fossil evidence place the origin of cichlid fishes long after Gondwanan rifting. Proceedings of the Royal Society B: Biological Sciences, 280(1770), 20131733.

137 Schofield, P. J., Loftus, W. F., Kobza, R. M., Cook, M. I., & Slone, D. H. (2010). Tolerance of nonindigenous cichlid fishes (*Cichlasoma urophthalmus*, *Hemichromis letourneuxi*) to low temperature: laboratory and field experiments in south Florida. Biological Invasions, 12(8), 2441-2457.

138 Van Wilgen, N. J., & Richardson, D. M. 2011. Is phylogenetic relatedness to native species important for the establishment of reptiles introduced to California and Florida? Diversity and Distributions, 17(1), 172-181.

139 Ma, C., Li, S. P., Pu, Z., Tan, J., Liu, M., Zhou, J., & Jiang, L. (2016). Different effects of invader–native phylogenetic relatedness on invasion success and impact: a meta-analysis of Darwin's naturalization hypothesis. Proceedings of the Royal Society B: Biological Sciences, 283(1838), 20160663.

140 Purvis, A. 1995. A composite estimate of primate phylogeny. Philosophical Transactions of the Royal Society of London. Series B: Biological Sciences, 348(1326), 405-421.

141 Frantz, L. A., Mullin, V. E., Pionnier-Capitan, M., Lebrasseur, O., Ollivier, M., Perri, A., & Tresset, A. (2016). Genomic and archaeological evidence suggest a dual origin of domestic dogs. Science, 352(6290), 1228-1231.

142 Thalmann, O., Shapiro, B., Cui, P., Schuenemann, V. J., Sawyer, S. K., Greenfield, D. L., & Napierala, H. (2013). Complete mitochondrial

genomes of ancient canids suggest a European origin of domestic dogs. Science, 342(6160), 871-874.

143 Vilà, C., Savolainen, P., Maldonado, J. E., Amorim, I. R., Rice, J. E., Honeycutt, R. L., & Wayne, R. K. (1997). Multiple and ancient origins of the domestic dog. Science, 276(5319), 1687-1689.

144 Freedman, A. H., Gronau, I., Schweizer, R. M., Ortega-Del Vecchyo, D., Han, E., Silva, P. M., & Beale, H. (2014). Genome sequencing highlights the dynamic early history of dogs. PLoS genetics, 10(1), e1004016.

145 Botigué, L. R., Song, S., Scheu, A., Gopalan, S., Pendleton, A. L., Oetjens, M., & Bobo, D. (2017). Ancient European dog genomes reveal continuity since the early Neolithic. Nature Communications, 8.

146 Parker, H. G., Dreger, D. L., Rimbault, M., Davis, B. W., Mullen, A. B., Carpintero-Ramirez, G., & Ostrander, E. A. (2017). Genomic Analyses Reveal the Influence of Geographic Origin, Migration, and Hybridization on Modern Dog Breed Development. Cell Reports, 19(4), 697-708.

147 Newby, J. (1997). The pact for survival: humans and their animal companions. ABC Books for the Australian Broadcasting Corporation.

148 Shahid, S.A., Xiao,Y., Khan, S., Feng, D., Johnson, G.S. and Ha, J. (2004). Sequence Diversity of the Canine Mitochondrial Genome. University of Missouri.

149 Webb, K. M., & Allard, M. W. (2009). Mitochondrial genome DNA analysis of the domestic dog: identifying informative SNPs outside of the control region. Journal of forensic sciences, 54(2), 275-288.

150 Imes, D.L. and Sacks, B.N. (2011). Identification of Single Nucleotide Polymorphisms within the mtDNA Genome of the Domestic Dog to Discriminate Individuals with Common HVI Haplotypes. University of California.

151 Angleby, H., Oskarsson, M., Pang, J., Zhang, Y. P., Leitner, T., Braham, C., .& Savolainen, P. (2014). Forensic Informativity of~ 3000 bp of Coding Sequence of Domestic Dog mtDNA. Journal of forensic sciences, 59(4), 898-908.

152 Pollinger, J. P., Lohmueller, K. E., Han, E., Parker, H. G., Quignon, P., Degenhardt, J. D., & Bryc, K. (2010). Genome-wide SNP and haplotype analyses reveal a rich history underlying dog domestication. Nature, 464(7290), 898.

153 Skoglund,P., Ersmark, E., Palkopoulou, E., & Dalén, L. (2015). Ancient wolf genome reveals an early divergence of domestic dog

ancestors and admixture into high-latitude breeds. Current Biology, 25(11), 1515-1519.

154 Lane, M. (2011). The moment Britain became an island. BBC History.

155 Darvill, T. (2010). Prehistoric Britain. Routledge.

156 Bradley, R. (2007). The prehistory of Britain and Ireland. Cambridge University Press.

157 Nqayi, Z. (2014). The commemoration of International Day for Biological Diversity. Department of Environmental Affairs, Republic of South Africa.

158 Hammond, N. (2008). "Flint hints at existence of Palaeolithic man in Ireland". The Times

159 Day, S. P. (1996). Dogs, deer, and diet at Star Carr: a reconsideration of C-isotope evidence from early

160 Mesolithic dog remains from the Vale of Pickering, Yorkshire, England. Journal of Archaeological Science, 23(5), 783-787.

161 Morgan, K. O. (Ed.). (2010). The Oxford History of Britain. OUP Oxford.

162 Salway, P. (2001). A history of Roman Britain. Oxford Paperbacks.

163 Yalden, D. (2010). The history of British mammals. A&C Black.

164 Maher, B. A. (2002). Uprooting the tree of life: a proposed theory has researchers debating life's origins--again. The Scientist, 16(18), 26-28.

165 Garamszegi, L. Z., & Gonzalez-Voyer, A. (2014). Working with the tree of life in comparative studies: How to build and tailor phylogenies to interspecific datasets. Modern Phylogenetic Comparative Methods and Their Application in Evolutionary Biology (pp. 19-48). Springer Berlin Heidelberg.

166 Short, R. (2009). King Canute and the wisdom of forest conservation. Nature, 462(7273), 567-567

167 Hannas, C. A. (1978). An analysis of hunting and sporting scenes portrayed in the decoration of glass, pottery, and porcelain wares in the Brunnier Collection.

168 Ireland, S. (2008). "Chapter 15: Government, Commerce and Society". Roman Britain: A Sourcebook. Routledge Sourcebooks for the Ancient World (3rd ed.). Taylor & Francis. p. 216. ISBN 9780415471770. OCLC 223811588.

169 Wideroff L, Vadaparampil ST, Greene MH, Taplin S, Olson L, Freedman AN. Hereditary breast/ovarian and colorectal cancer genetics knowledge in a national sample of US physicians. Journal of Medical Genetics. 2005;42(10):749-755.

170 Burke W, Daly M, Garber J, Botkin J, Kahn MJE, Lynch P, Thomson E. Recommendations for follow-up care of individuals with an inherited predisposition to cancer: II. BRCA1 and BRCA2. Jama. 1997;277(12):997-1003.

171 Vogelstein B, Kinzler KW. Cancer genes and the pathways they control. Nature Medicine. 2004;10(8):789.

172 Fan S, Meng Q, Auborn K, Carter T, Rosen EM. BRCA1 and BRCA2 as molecular targets for phytochemicals indole-3-carbinol and genistein in breast and prostate cancer cells. British Journal of Cancer. 2006;94(3):407-426.

173 Gayther SA, de Foy KA, Harrington P, Pharoah P, Dunsmuir WD, Edwards S M, Ford D. The frequency of germ-line mutations in the breast cancer predisposition genes BRCA1 and BRCA2 in familial prostate cancer. Cancer Research. 2000;60(16):4513-4518.

174 Mitra A, Fisher C, Foster CS, Jameson C, Barbachanno Y, Bartlett, Easton D. Prostate cancer in male BRCA1 and BRCA2 mutation carriers has a more aggressive phenotype. British Journal of Cancer. 2008;98(2):502-507.

175 King MC, Marks JH, Mandell JB. Breast and ovarian cancer risks due to inherited mutations in BRCA1 and BRCA2. Science. 2003;302(5645):643-646.

176 Puente XS, Velasco G, GutiérrezFernández A, Bertranpetit J, King MC, López-Otín C. Comparative analysis of cancer genes in the human and chimpanzee genomes. BMC Genomics. 2006;7(1):15.

177 Sharma S, Kelly TK, Jones PA. Epigenetics in cancer. Carcinogenesis. 2010;31(1):27-36.

178 Lou DI, McBee RM, Le UQ, Stone AC, Wilkerson GK, Demogines AM, Sawyer SL. Rapid evolution of BRCA1 and BRCA2 in humans and other primates. BMC Evolutionary Biology. 2014;14(1): 155.

179 Huttley GA, Easteal S, Southey MC, Tesoriero A, Giles GG, McCredie MR, Venter DJ. Adaptive evolution of the tumour suppressor BRCA1 in humans and chimpanzees. Nature Genetics. 2000; 25(4):410-413.

180 Pavlicek A, Noskov VN, Kouprina N, Barrett JC, Jurka J, Larionov V. Evolution of the tumor suppressor BRCA1 locus in primates: Implications for cancer predisposition. Human Molecular Genetics. 2004;13(22):2737-2751.

181 Orban TI, Olah E. Emerging roles of BRCA1 alternative splicing. Molecular Pathology. 2003;56(4):191.

182 Roncaglioni MC, Falanga A, D'Alessandro AP, Alessio MG, Casali B, Donati MB. Evidence of a warfarin-sensitive cancer procoagulant in V2 carcinoma. Haematologica. 1989;74(2):143-147.

183 Holt JT, Thompson ME, Szabo C, Robinson-Benion C, Arteaga CL, King M.C, Jensen RA. Growth retardation and tumour inhibition by BRCA1. Nature genetics. 1996;12(3):298.

184 Huang YW. Association of BRCA1/2 mutations with ovarian cancer prognosis: An updated meta-analysis. Medicine. 2018;97(2):e9380.

185 Orr HA. The distribution of fitness effects among beneficial mutations. Genetics. 2003;163(4):1519-1526.

186 Perfeito L, Fernandes L, Mota C, Gordo I. Adaptive mutations in bacteria: High rate and small effects. Science. 2007; 317(5839):813-815.

187 Charlesworth B, Morgan MT, Charlesworth D. The effect of deleterious mutations on neutral molecular variation. Genetics. 1993;134(4):1289-1303.

188 Foulkes WD, Knoppers BM, Turnbull C. Population genetic testing for cancer susceptibility: Founder mutations to genomes. Nature Reviews Clinical Oncology. 2016;13(1):41.

189 Banerji S, Cibulskis K, Rangel-Escareno C, Brown KK, Carter SL, Frederick AM, Cortes ML. Sequence analysis of mutations and translocations across breast cancer subtypes. Nature. 2012; 486(7403):405-409.

190 Charlesworth B, Morgan MT, Charlesworth D. The effect of deleterious mutations on neutral molecular variation. Genetics. 1993;134(4):1289-1303.

191 Susan A. Johnston, Religion, Myth, and Magic: The Anthropology of Religion—A Course Guide (Recorded Books, LLC, 2009).

192 David Adams Leeming and Margaret Adams Leeming, A Dictionary of Creation Myths, Oxford Reference Online Ed. (Oxford University Press, 2009).

193 Dawkins, The Greatest Show on Earth: The Evidence for Evolution (United Kingdom: Free Press, Transworld, 2009).

194 US Geological Survey. (2020). "Age of the Earth."

195 Origins of Life and Evolution of the Biosphere (Netherlands: Kluwer Academic Publishers, 2000), 30: 107-112. doi:10.1023/A:1006746205180.

196 C. Darwin, On the Origin of Species (United Kingdom: John Murray, 1859).

197 Douglas J. Futuyma, Evolution (Sunderland, Massachusetts: Sinauer Associates, Inc., 2005).

198 J. W. Schopf, Cradle of life: the discovery of Earth's earliest fossils. (Princeton, 1999).

199 Woese, C. (1998). The universal ancestor. Proceedings of the national academy of Sciences, 95(12), 6854-6859.

200 D. L. Theobald, "A formal test of the theory of universal common ancestry," Nature (2010).

201 Hodge, T., Jamie, M., & Cope, T. V. (2000). A myosin family tree. Journal of cell science, 113(19), 3353-3354.

202 Kimura, M. (1968). Evolutionary rate at the molecular level. Nature, 217(5129), 624-626.

203 Turner, C. B., Blount, Z. D., Mitchell, D. H., & Lenski, R. E. (2015). Evolution and coexistence in response to a key innovation in a long-term evolution experiment with Escherichia coli. bioRxiv, 020958.

204 Ross, J., & Adkins, R. (2018). Biodiversity and Environmental Conservation. Scientific e-Resources.

205 Peppe, D. J., & Deino, A. L. (2013). Dating rocks and fossils using geologic methods. Nature Education Knowledge, 4(10), 1.

206 Ayala, F. J. (2008). Science, evolution, and creationism.

207 Emerson, A. E. (1938). Termite nests--a study of the phylogeny of behavior. Ecological Monographs, 8(2), 247-284.

208 I. Letunic and P. Bork, "Interactive Tree Of Life (iTOL): an online tool for phylogenetic tree display and annotation," (Pubmed, 2007). Bioinformatics.

209 Theobald, D. L. (2010). A formal test of the theory of universal common ancestry. Nature, 465(7295), 219-222.

210 D. M. Mount, Bioinformatics: Sequence and Genome Analysis (second ed.) (Cold Spring Harbor, NY: Cold Spring Harbor Laboratory Press, 2004).

211 Barbagli, F. (2009). In Retrospect: The earliest picture of evolution?. Nature, 462(7271), 289-289.

212 Fan, Y., Linardopoulou, E., Friedman, C., Williams, E., & Trask, B. J. (2002). Genomic structure and evolution of the ancestral chromosome fusion site in 2q13–2q14. 1 and paralogous regions on other human chromosomes. Genome Research, 12(11), 1651-1662.

213 Ludwig, L. S., Lareau, C. A., Ulirsch, J. C., Christian, E., Muus, C., Li, L. H., & Law, T. (2019). Lineage tracing in humans enabled

by mitochondrial mutations and single-cell genomics. Cell, 176(6), 1325-1339.

214 Nielsen, R., Akey, J. M., Jakobsson, M., Pritchard, J. K., Tishkoff, S., & Willerslev, E. (2017). Tracing the peopling of the world through genomics. Nature, 541(7637), 302-310.

215 Kolb, A. W., Ané, C., & Brandt, C. R. (2013). Using HSV-1 genome phylogenetics to track past human migrations. PloS one, 8(10), e76267.

216 Sousa, V., & Hey, J. (2013). Understanding the origin of species with genome-scale data: modelling gene flow. Nature Reviews Genetics, 14(6), 404-414.

217 Barbujani, Guido & Colonna, Vincenza. (2013). Human Populations: Origins and Evolution. 10.1002/9780470015902.a0001794.pub3.

218 Numbers, R. L. (2006). The creationists: from scientific creationism to intelligent design (No. 33). Harvard University Press.

219 Dawkins, R. (2016). The selfish gene. Oxford university press.